DAVID S. RYAN
2118 OAK MEADOW CIRCLE
SOUTH DAYTONA BEACH, FL 32119

MANAGING LARGE RESEARCH AND DEVELOPMENT PROGRAMS

SUNY Series in Administrative Systems
Mariann Jelinek, General Editor

Managing Large Research and Development Programs

Henry W. Lane
Rodney G. Beddows
Paul R. Lawrence

STATE UNIVERSITY OF NEW YORK PRESS
Albany

Published by
State University of New York Press, Albany

© 1981 State University of New York

All rights reserved

Printed in the United States of America

No part of this book may be used or reproduced
in any manner whatsoever without written permission
except in the case of brief quotations embodied in
critical articles and reviews

For information, address State University of New York
Press, State University Plaza, Albany, N.Y., 12246

Library of Congress Cataloging in Publication Data

Lane, Henry W., 1942–
  Managing large research and development programs.

  Bibliography: p. 157
  Includes index.
  1. Research, Industrial—Management. I. Beddows,
Rodney G., 1944–   II. Lawrence, Paul R.
III. Title.
T175.L28   607'.2'68   81–849
ISBN 0–87395–473–4   AACR2
ISBN 0–87395–474–2 (pbk.)

9 8 7 6 5 4

Contents

Acknowledgments x
List of Illustrations viii
List of Tables ix

1. R & D MANAGEMENT: OLD VIEWS AND NEW REALITIES
   Introduction 1
   Some Viewpoints on R & D Management 2
   Research on R & D 3
   Focus at the Technological Core 3
   The Context of R & D 4
   Political Intervention in R & D: A New Reality 5
   Background of the Research 6
   The Two Organizations 8
   The Program Sample 9
   Summary 13

2. THE POLITICAL CONTEXT OF RESEARCH AND DEVELOPMENT
   Introduction 16
   A Dominated Program: Sickle Cell Anemia 18
     The Stimulus for a Program 19
     Response to the President's Initiative 20
     Further Responses 21
     Results 22
     Commentary on the Sickle Cell Anemia Program 23
   Shifts in the Relationship with the Political Environment 25
   Destabilization and Readaptation: High Capacity Mobile
     Telephone System 25
   The Nature of the Political Environment 28
   Patterns of Mutual Adaptation 29

3. THE TECHNICAL LOGIC OF RESEARCH AND DEVELOPMENT
   The Technical Logic of Research and Development Programs 34

The Development of Genetics Knowledge 36
Technical Logic of Biomedical Research 38
Comparing the Technical Logic of Biomedical and
  Communications Research 41
The Millimeter Waveguide Program 44
The First Exploratory Development Program 46
The Second Exploratory Development Program 47
Manufacturing Waveguide for Testing 47
Installation Techniques 48
Field Evaluation Test 49
Waveguide Put "On Hold" Once Again 49
The Technical Logic of Millimeter Waveguide 49

4. LINKING THE POLITICAL AND TECHNICAL ENVIRONMENTS: TWO UNSUCCESSFUL PROGRAMS
   The Artificial Heart Program 54
      The Master Plan 56
      Response to the Master Plan 57
      The Political Environment of the Artificial Heart Program 59
      The Technical Logic of the Artificial Heart Program 60
   Comparing the Waveguide and Artificial Heart Cases 63
   Organizing to Fight Cancer 66
   Emergence of the Chemotherapy Program 67
   The Political and Technical Logics of Chemotherapy 69

5. LINKING THE POLITICAL AND TECHNICAL ENVIRONMENTS: A SUCCESSFUL PROGRAM
   Sudden Infant Death Syndrome (SIDS) 76
   Activity in the Political Environment 77
   Initiating a SIDS Program and Sizing-Up the State of
      Knowledge 79
   Managing the Political Environment 81
      Congress Becomes Involved with SIDS 81
      Interacting with Parent Groups 83
      Interacting with the Administration 84
      Commentary on the SIDS Program's Interactions 85
   Contrasting the SIDS, Artificial Heart, and Sickle
      Cell Programs 86
   The Technical Logic of SIDS 87
   Administrative Mechanisms Supporting the Technical Logic 89
      Frank Hastings—Technique Champion 94
      Bill Warters—Program Champion 95

Contents

6. ORGANIZING THE TECHNICAL LOGIC
   The Bell Laboratories Organization   100
   Managing Knowledge Transfer   102
   Discovery Activities at Crawford Hill (Stages 1–3)   103
      The Radio Research Lab   105
      Funding and Project Selection   106
      Communicating Results   107
   Exploratory Development (Stage 4)   108
   Development (Stage 5)   109
      The Branch Lab at the Merrimack Valley Works   110
      The D4 Program at the Merrimack Valley Works   112
      The No. 4 Electronic Switching Systems (No. 4 ESS) Program
      Organizational Decisions   114
      Implementation   119
   Linking Task and Organization   119

7. ADAPTATION TO CHANGES IN THE TECHNICAL AND POLITICAL ENVIRONMENTS
   Origin of Genetics Research Activity at NIH   124
   The Early Genetics Program   125
   The Changing Context of the Genetics Program   128
   A Second Mode of Adaptation   129
   Managing the Genetics Program After 1972   131
      The Program's Role in Science   132
      The Program's Role at the Boundary of Science and Society   134

8. MANAGING LARGE R & D PROGRAMS: SUMMARY AND CONCLUSIONS
   The R & D Management Model   140
   Managing Boundaries   142
   Creating a Synthesis   145
      Dual Advocacy   145
   Establishing Criteria   147
   Switching: The Process of Organizational Learning   149
   Using the Model   151

   Bibliography   157
   Index   163

Figures

1. Program Selection Matrix   10
2. Research Framework   15
3. Chapter 2 Focus   16
4. Chapter 3 Focus   34
5. Technical Logic of Communication Systems Research   41
6. Sequential Research Logic   43
7. Empirical Research Logic   43
8. Stages in the R & D Process   45
9. Chapter 4 Focus   54
10. The Vicious Circle of the Chemotherapy Program   70
11. The Task Logic of the Artificial Heart Program   71
12. The Task Logic of the Chemotherapy Program   72
13. Chapter 5 Focus   75
14. The Vicious Circle Surrounding SIDS   78
15. Stages in the R & D Process   97
16. Chapter 6 Focus   99
17. Bell Labs Organization (1976)   101
18. The Technical Logic of R & D   102
19. A Continuum of Uncertainties Facing R & D Programs   102
20. The Crossover   103
21. Transfer of Design Information   113
22. PECC Interfaces—D4 Program   114
23. Bell Labs and Western Electric Organizations Involved in the No. 4 ESS Program   118
24. The One System Concept   121
25. Chapter 7 Focus   124
26. Genetics Program Boundary Placement   136
27. R & D Management Model   141
28. Tension and Learning   144
29. Variety and Learning   151

Tables

1. Technical Logic of Biomedical Research   40
2. Millimeter Waveguide and Artificial Heart Comparison   64–65
3. OPS Form for the SIDS Program   92–93
4. Early Genetics Program   127
5. Genetics Program—Post 1972   137
6. Comparison of Two Adaptive Modes   138
7. Some Selected Heuristics   152

Acknowledgments

The research reported in this book was funded by the National Institutes of Health under Contract No. 1–OD–6–2109. In early 1975 the director of NIH, Dr. Robert Stone, approached Paul Lawrence with the idea of doing a systematic study of research management at NIH. Dr. Stone, in his previous roles as dean of the Medical School, University of New Mexico, and visting scholar at the Sloan School of Management at MIT, had collaborated with Lawrence in a comparative study of medical schools. Dr. Stone's interest in the project went beyond a general interest in the subject of management or a desire to extend a prior relationship. NIH, at that time, was under pressure to make more use of "modern business management methods" and anticipated continuing pressure in the future. Dr. Stone decided to take the initiative by launching this study.

We are grateful to Dr. Stone for starting the project; to Bruce Carson and other members of the Office of the Director of NIH who contributed to its progress and completion; and to the many people in staff and management positions in the various Institutes who willingly shared their time and knowledge with us.

We also thank the executives of Bell Telephone Laboratories and Western Electric for agreeing to participate in this research, and the people in these organizations who were most cooperative and from whom we learned a great deal. W. M. "Red" Mackay, in particular, was very helpful during the project.

To everyone in NIH and the Bell System with whom we dealt, we express our appreciation for their assistance.

There were other important sources of support: The Plan for Excellence at the School of Business Administration, University of Western Ontario, which funded Henry Lane to continue working on this project; members of the Lane family who lived with this project for a long time; MaryAnn Lowry who edited drafts of the chapters; Harvey Kolodny who provided continual encouragement; and Mariann Jelinek who, in addition to providing encouragement, worked with us to help clarify what we were saying.

H.W.L., R.G.B., P.R.L.

Chapter 1

R & D Management: Old Views and New Realities

Introduction

Our society sorely needs the innovations that are the intended fruit of research and development. While social commentators have differences of opinion about man's global prospects, they tend to agree that a major crunch is coming between our limited resources and our rising expectations. We may not agree on the size of our resource reserves, but all must recognize their finite nature. The obvious source of relief from these conflicting realities is widespread innovation, both technical and social. The innovative process that has been essential for biological survival is now the critical element for our survival as civilized societies.

Yet, there is reason to question the effectiveness of our current management of R & D resources. The amount of resources going into R & D is large by any standard and it is increasing. In the United States alone the annual amount was estimated at $47 billion in 1977.[1] Though comparable figures are not available on a world basis, the amount is probably twice that of the United States and increasing at a faster rate.[2] With confidence that these resources were being managed with minimal waste, there would be little cause for concern. But such is not the case! There are just too many indications to the contrary. Of course, any reasonable observer would be quick to acknowledge that, since R & D involves exploring uncertainties, even under ideal management there will always be resources used for false starts and dead-end studies. What is disturbing is the evidence that such legitimate initial explorations mushroom too often into large scale R & D efforts on what eventually must be labeled hopeless causes. These constitute the enormous areas of waste where it is reasonable to expect better management to make a significant difference. The examples that have caught the public's attention—such as the C5A, the SST, and the F-111—may be just the tip of

---

[1] *Business Week*, July 3, 1978, pp. 58 and 47 ($18B by private industry—$29B by federal government).
[2] *Business Week*, July 3, 1978, p. 47.

the iceberg. Similar situations in private industry are less likely to reach the press, but insiders exchange comparable stories of large, long-term R & D programs that have come to naught.

When news of such mistakes gets into the press, the public often visualizes a management group that either wilfully misdirects a program for self-centered reasons, or misleads it out of gross stupidity. Nothing in our research would support such conclusions. To the contrary, what we saw were bright, hard-working, well-intentioned people who in spite of excellent, good-faith efforts could all too easily get trapped into supporting large programs long after they could be seen to be nearly hopeless from another perspective. Why does it happen? One reason it happens is our lack of knowledge about the R & D process and its management. Even the brightest and best R & D managers must move across largely unchartered seas. Under these conditions it seems more sensible to respond to episodes of wrecks and wastage by doing more charting rather than by cursing management for corruption, negligence or stupidity.

Reasons for concern about improving R & D management are not hard to find. The purpose of this research was to contribute to the improvement of the research and development management process.

Some Viewpoints on R & D Management

One indicator of inadequate guidance for senior R & D managers is the prevalence of faddish swings in management practice. In the post-World War II era, when confidence in the fruits of research was high, conventional wisdom held strongly that the best R & D laboratories were the least managed ones. This idea was carried so far that some otherwise sensible executives believed the best way to manage R & D was to hire an assortment of well-trained scientists, put them in a well-stocked laboratory in some remote pastoral spot and eventually market the breakthroughs certain to emerge. R & D seemed to be a straight-forward process of hiring some geniuses and leaving them alone.

When this approach gradually became discredited through painful experience, management style swung the other way. In some instances, an extreme of calling for weekly written progress reports, PERT charts to pre-plan every step of the inquiry, advance budgeting to the dollar, and more emerged. The attitude of "leave them alone" was being replaced with one of "hold them accountable." Research scientists, accustomed to autonomy and abundant resources, undoubtedly found it difficult to embrace management's new approach and found themselves on the defensive. Such swings and widely divergent beliefs about how best to manage research are clear evidence that we need to learn more about the process if we are to improve R & D performance.

Research on R & D

Academics have shown a keen interest in R & D. The literature on the topic, while not insignificant in amount and quality, is not a coherent body of knowledge. Although a number of issues has been investigated, other equally important aspects have not been addressed adequately. Particularly apparent is a gap in knowledge about the management of large, complex research programs and, even more specifically, about the interaction between the organizations in which they occur and an increasingly volatile political environment.

Focus at the Technological Core

A large body of literature has the R & D laboratory, or technological core, of the organization as its focus and deals with various human and environmental aspects of it.

Pelz and Andrews (1966) examined such topics as the motivations, satisfaction, creativity, diversity, and dedication of scientists, as well as the prevailing social dynamics within the R & D groups. Their objective was to link characteristics of individuals and the climate of the labs with individual effectiveness. Allen's (1972) work was a significant step toward understanding behavior in science laboratories. The concept of the "technological gatekeeper" and the importance of that role was a major contribution. Barnowe (1975) investigated the effect of leadership on a laboratory and found that 18 percent of its scientific and research contribution could be attributed to the technical skill, supportiveness, and participative orientation of the leaders.

There also had been research into the value orientation of scientists; that is, into their different attachments, loyalties, and reference groups. The work of Gouldner (1951), and Kornhauser and Hagstrom (1962) focused on one dimension along which scientists varied which they labeled "local-cosmopolitan." Local scientists were committed primarily to the organization for which they worked and saw their professional careers as organization-based; cosmopolitan scientists were more committed to their specialization and research achievements, and related their careers to general professional recognition and to external reference groups. Friedlander (1970) investigated this dimension in greater depth, identifying three independent factors or poles which he called research, professional, and local orientation.

Still at the technological core, the focus shifts from the people to the research projects and a body of literature concerned with project selection and associated resource allocation decisions. There appear to be two major approaches to this area of interest: studies leading to prescriptive statements based on anecdotal evidence supplied by executives, and empirical studies utilizing statistical methods. Both emphasize the utility of formal tools for analysis and decision making. Twiss' (1974) prescription for research productivity is to develop more logical and scientific techniques for project selection, budgeting, control, and evaluation. Mansfield (1971) reported a series of studies in two research-intensive industries: chemicals and petroleum. Many of the quantitative methods suggested for project choice are adaptations of capital budgeting techniques. Mansfield discovered that about three-quarters of the laboratories were using such techniques but only in a limited way. Intuition and hunch remained major determinants of the data such as estimates of commercial success or market size that were inputs to the techniques. Laboratory administrators' estimates for such things as development time and manpower requirements also were subjective and exhibited enormous variation. A general conclusion of researchers in this area is that more rigorous use of formal techniques forces people to think more carefully about relevant factors and thus improves project evaluation and selection.

Many models of the R & D process have described a sequence of stages from "basic" through "development." These models attempt to capture differences in the state of knowledge, orientation of the researchers, and/or the immediacy of commercial application. Once the notion of stages had been established, researchers could ask the question about how knowledge was transferred from the beginning of the process (basic research) through to the end (production). Jermakowicz's (1978) research looked at organizational structures best integrating the R & D and production activities. He found that project-type organizations produced products with the most originality while matrix structures seemed to generate the greatest quantity of new products.

There is no question that having the right people with the right orientations, establishing the right laboratory conditions and climate, using the right leadership style, choosing the right project, and using the right structure all contribute to R & D performance. As important as these components are, however, they address issues primarily on the technical side of an organization. Research and development, particularly the large programs examined in this study, take place in large organizations which are embedded in complex social, economic and political environments. We believe it is important to understand the research process and the organization of which it is a part within this large context.

The Context of R & D

There is research that considers factors beyond the organization's boundaries. Most of it has concentrated on economic and market dimensions.

Studies of innovations and the innovation process have taken a dynamic approach to the stages concept of R & D. In particular, they recognized the importance of the internal interfaces between knowledge development and application and between the R & D process and the marketplace. Probably the most extensive study of industrial innovation, project SAPPHO (1972) investigated more than forty matched pairs of successful and unsuccessful innovations. Two of the primary success factors were understanding users' needs and emphasizing marketing.

Economists have devoted most of their attention in this area to the relationship between the economic structure of an industry, and research and development funding. This work generally has attempted to explain innovativeness by linking two important aspects of industrial structure: the size of firms and the degree of competition in a given industry or market related to the level of R & D expenditure.

Finally, we would mention the work of researchers who examined the social and cultural conditions fundamental to the development of science as a social institution. Merton (1968, 1972) identified the values and norms governing the behavior of those whose allegiance is to the scientific community. He commented on the increasing interdependency of science and its social environment that contributes to competition and conflict between these differing value systems: the traditional autonomy of the scientific community challenged by government. This type of penetrating but broad sociological commentary heralds the emergence of a new reality: increased political intervention into R & D. This issue lacks specific treatment in the literature, although the organizations we studied were very much concerned by it.

Political Intervention in R & D: A New Reality

Historically, society probably has relied more on the gifted individual, the inventor or the entrepreneur, for important contributions through new technologies. In today's world, R & D work is conducted mostly as large-scale efforts embedded in even larger organizations. Even though we will continue to need the individual inventor and the entrepreneur, the major emphasis has shifted to the management of large, highly organized R & D efforts. Essentially, we are forced to turn to our large institutions, both private and public, as major sources of technological and organizational innovation at the very time when the need is especially acute. And large institutions do not have all that impressive a record regarding major innovations and adaptations.

It is difficult for these large organizations to escape the notice of the public or government. For example, the auto, chemical, pharmaceutical, and nuclear power industries all are feeling the impact of social pressure and government intrusion into their technological processes and decision-making. Whether or not good reasons for this intervention exist can be debated, but the reality is directly observable. One needs only to look at the auto industry to see how the political process has set R & D agendas for years to come, with regulations governing emission, mileage, and safety standards.

The relevant issues for senior R & D administrators go beyond behavioral conditions in the laboratory, market studies, and detailed decision models. These people should be thinking about the critical social and political forces that can govern a firm's policy process and reduce its decision-making latitude.

The importance of these social and political factors was apparent immediately at the National Institutes of Health (NIH) and also surfaced in the activities of AT&T—the two organizations we studied. NIH is a government-supported institution and, as such, has to relate on a daily and long-term basis with political elements and external special-interest groups. Both groups may have had norms quite divergent from those of the scientific community, yet their support, in many cases, was necessary for the success of a research program. The technological core of AT&T was perhaps less open and less susceptible to direct intervention than was that of NIH, but it also had representatives of the political system, such as regulatory agencies and the Justice Department, with which to deal.

What we found lacking in the R & D literature was an appreciation of the specific ways in which external political forces affected the R & D process, and of how organizations respond to political incursion, and manage the interface between their technical and political environments. Given a potentially wide divergence between the norms and goals of a scientific community and the political system, the opportunity exists for development of either a beneficial creative tension or a destructive pathological relationship. Extending knowledge on this interface of the R & D process required, first, a description of the relationships between research organizations and external groups and, second, an analysis of those situations to determine what managerial tools could be used to facilitate beneficial inter-institutional relationships.

Background of the Research

By early 1975, the National Institutes of Health was under pressure from the Department of Health, Education and Welfare (HEW) and the Office of Management and Budget (OMB) to make more use of "modern management methods." These agencies were concerned that NIH's autonomy was resulting in excessively expensive basic biomedical research and in research programs unrelated to the nation's most pressing health needs. The creation of the President's Biomedical Research Panel, whose mandate was to examine policy, organizational and operational issues related to NIH-supported research, added to this pressure. NIH was spending approximately $2 billion—a significant amount of money—on biomedical research. Inflation was a continuing problem; research costs were escalating; multiple demands were being placed on a limited pool of funds; and pressure for targeting research to major health problems was increasing. With the mounting desire to get more "bang for the buck," the pressure exerted on NIH for improved administrative and fiscal efficiency was understandable.

The most controversial issue was identifying the "best way" to manage research. People holding one extreme orientation to the issue believed that it was almost impossible to control and evaluate in-process scientific research, and that any planning of research should be done by scientists using scientific criteria. Scientific inquiry was a creative task requiring time and autonomy. Proponents at the other extreme argued that the best way to ensure progress was through tight control systems and targeted research efforts maximizing immediate return on the invested tax dollar. The latter position had gained strength, fueled by rising costs for research and increased social and political pressure to solve pressing public health problems.

A change in emphasis away from basic research and toward applied research had been apparent over the years, and was viewed with some concern within NIH. A former director of NIH expressed a common criticism of this trend:

> This change would appear to reflect a desire to emphasize short- and middle-term objectives of a specific nature in preference to a balanced mix of those with middle- and long-range objectives, some of the latter being of a more general nature. Such tendency to over-emphasize that which is immediately practical, can tend, even in the short view, to limit the likelihood of an ultimate solution of the more important problems of medicine within any reasonable timeframe.[3]

The two contrasting attitudes on how best to manage research can be summarized as follows:

FOCUS: Society's Problems vs. Science's Problems

ORIENTATION: Maximize Output (Applied Research) vs. Maximize Knowledge (Basic Research)

TIME HORIZON: Short Term vs. Long Term

LOCUS OF CONTROL: Managers/Administrators Evaluate (External) vs. Scientists Evaluate (Internal)

DEGREE OF CONTROL: Tight Control vs. Autonomy

---

[3]Speech by Dr. James Shannon on the 10th anniversary of the National Institute of General Medical Sciences. HEW Publication NIH 74-274; May 21, 1973.

The pressure for change was not solely from outside NIH, however. The NIH director in 1975 also was interested in improving the Institutes' management systems and program evaluation procedures. He anticipated increased pressure on NIH to improve its research management activities in the future. Perhaps on the theory that a good offense is the best defense, the director initiated this study of the historically-determined management practices at NIH. His hypothesis, that HEW's and OMB's version of "modern business management methods" which emphasized tight budgets, schedules, and control might be useful for only a limited number of NIH programs, was shared by the researchers.

We also believed that comparison was an important consideration. In organizational research, as in many of the "hard" sciences, the comparative approach has added valuable perspectives in understanding and explaining complex phenomena. Thus, the study was designed to provide both internal and external comparisons. Chosen for external comparison were programs of Bell Telephone Laboratories which was engaged in a wide spectrum of basic, applied, and development research activities.

It is clear why NIH wanted to conduct this study. The political system was becoming more actively involved in affairs NIH scientists traditionally had considered their exclusive domain. But why did Bell want to be involved? Our answer can be only speculative, but it was obvious that Bell, too, was feeling considerable political pressure, from the government, existing and potential competitors, and private, special interest groups. The two organizations shared a similar problem: each was a prestigious and successful R & D institution, staffed by talented people who believed in the basic soundness of their organizational arrangements and management methods but who were having trouble communicating about them with the political system. Both believed that this study could be useful in mapping out their critical organizational processes and in providing a more systematic way of thinking and talking about them.

The Two Organizations

We studied programs in two very large, complex, highly visible and unique institutions. NIH, with a budget of approximately $2 billion, is perhaps the world's focal point for biomedical research support. Its organization is as complex as its scientific mission. It is a non-profit organization that can be characterized as polycentric and relatively open to political pressure. Each Institute is essentially a center of power revolving around a central administrative and policy core. The programs we studied were contained within these semi-autonomous institutes. AT&T, on the other hand, is profit-oriented and is a much more highly integrated organization displaying more central coordination and control than NIH. Bell Laboratories, the primary locus of research and development activity in the Bell System, is, in its own right, unique in size and accomplishments. Its budget exceeds $600 million and it employs more than 16,000 people including some Nobel Laureates.

Another difference that was dramatic, but not extremely important for our purposes, was the number of personnel directly employed. All the NIH programs had very small staffs compared to AT&T programs, for the simple reason that the NIH program staff only provided senior management or top level guidance to the program; almost all of the technical work and its direct supervision was "bought" elsewhere by grants and contracts. In contrast, the AT&T efforts were all "in house."

Within these organizations, our focal programs ranged widely in size and maturity. We studied programs whose origins go back 40 years and others which are less than a decade old; programs whose budget ran into hundreds of millions of dollars and those costing about three million dollars per year; those with a basic focus as well as development programs.

Given the uniqueness in the institutions we were examining, we think it would have been inappropriate to have placed high demands on perfectly matched samples. Even if we had achieved a representative sample, we would have been open to the fundamental criticism that NIH and Bell are not representative territories for the research. This criticism could be valid if one is concerned with structural comparability alone. On the other hand, they are exciting and interesting domains to examine simply because they are large and have a very long and successful history in achieving research and development objectives. In this sense they fit a research strategy whereby much can be learned through observation of the unique and unusual rather than the standard.

These issues of comparability may be summarized as unique vs. similar, public vs. private, non-profit vs. profit, and structure vs. process.

We compared two very different sites and found some very similar processes across the gap of "incomparability." The focus was the management process and our findings, we believe, can increase understanding of the management of large research development programs, particularly in the balancing of technical and political considerations. In most instances, these processes took place at a program level with NIH; and while many of these issues also were addressed at the program level at AT&T, some were addressed at the corporate level.

The Program Sample

We proposed that NIH, which funded the study, identify a sample of its programs including both relatively high and low performers operating under relatively certain and uncertain technical and environmental conditions. NIH accepted this plan and the proposed comparison with an industrial R & D organization well in advance of Bell's agreeing to take part in the research.

During the preliminary stages of the research-project definition, NIH staff members carefully compiled a list of 25 programs from which the sample would be drawn. These programs, generally representative of the type found at NIH, were arranged on a continuum according to their degree of certainty or unpredictability:

a) *At the uncertain/unpredictable end* of this continuum was the classic basic laboratory research with investigators determining what needed to be done and program managers having more knowledge of what needed to be explored than where the exploration would lead.

b) *Certainty/predictability of goals*. This point represented a program where the manager knew where he wanted to go, but was not sure how to get there.

c) *Certainty/predictability of goals and approaches*. The goals and approaches to the problem were known, but uncertainty existed about attainability of the goals given existing knowledge.

d) *Certainty/predictability of goals, approaches, and results*. Only one project fit this characteristic and it was part of a larger effort. NIH staff members felt this extreme of the continuum should be represented even though a discrete program could not be found to fit it. They believed that the nature of the NIH mission made it unlikely that anything but a project or sub-program would meet this degree of certainty.

These programs then were rated by twenty-one anonymous scientists at NIH using a scale of 1 (poorest performance) to 7 (best performance). The rating technique and evaluators were chosen by the NIH staff members. Programs rated 5 and above were considered to be higher performers and those rated 4 and below, lower performers.

We used a decision rule of 75 percent response rate (the evaluators were not all familiar with all the programs) and 80 percent agreement of the performance level to provide the initial program selection matrix shown in Figure 1.

FIGURE 1
PROGRAM SELECTION MATRIX

		Certainty	
		Uncertain	Certain
Performance	Higher	Genetics	Hypertension
	Lower	Sickle Cell	Artificial Heart

The research began with a pilot study of one program. The purpose was to offer the researchers an in-depth look at a program as a way of learning the

R & D Management: Old Views and New Realities 11

complexities of NIH prior to final sample selection. The Sudden Infant Death Syndrome (SIDS) program was chosen since NIH representatives believed the program manager had given considerable thought to the management process and therefore SIDS might be an illuminating place to begin studying the management process within NIH. An additional factor in its suggested research value was the source of this program. The SIDS program had been externally initiated by Congress, without additional funding—apparently a test for any manager.

After a week of interviews with staff of the SIDS Program and the National Institute of Child Health and Human Development, and review of documents, the program sample was finalized. From our discussions with program and other Institute personnel and staff members in the Office of the Director, and from a review of the R & D literature, the potential significance of widely divergent norms and values emerged strongly. It became apparent that the perceived "success" of a program might be related to whether it was internally initiated by NIH or the scientific community, or whether it was externally initiated by Congress or the Executive Branch. We were unsure as to what characteristics of a program were being evaluated. Was a "basic" program receiving a higher rating than an "applied" one, for example?

Again from our interviews, we came to question the certainty dimension in view of divergent orientations. It seemed clear that even the most "applied" effort could face a significant amount of uncertainty or unpredictability, although perhaps of an administrative, social or political rather than scientific type. And thus the scientific community may see it as relatively certain since the "hardest part" is behind.

We learned that, while the senior people at NIH generally agreed on the relative performance of programs, they differed on the certainty/uncertainty of these programs' respective task environments. These programs had extensive histories (as much as 20 years) and often had gone through a number of stages with major organizational shifts. These two discoveries were linked: this history of stages and shifts made it very difficult to rate task and environmental uncertainty. It seemed task and environmental realities had changed too much within a given program history for different observers to reach any reasonable consensus about their ratings. The programs finally selected for study therefore were chosen to provide a range of performance, but we could not be confident of their differences in terms of environmental certainty.

Finally we were concerned that since somewhat polar orientations were emerging as potentially significant forces, the Cancer Institute, oldest, largest, and most visible of the Institutes, was not represented. This Institute had a unique relationship with NIH. Its direct funding from Congress gave it a great deal of autonomy from the Office of the Director of NIH.

Even as we were forced to face up to the impracticality of our original design, we became intrigued with the idea of broadening our study focus to understand better the history of these programs and the nature of the shifts or adaptations that were thrust upon them—sometimes by their own technical success at an earlier R & D stage, but also by political and special-interest-group pressures. We saw a need to extend our research frame of reference in order to consider these factors which clearly were critical to the program results. The emerging challenge was to try to understand and relate the two key forces acting on each of these programs—the moving state of technical knowledge, and the needs and pressures of the wider social and political world providing the resources. We saw a critical need to relate these very different and crucial realities with which R & D senior managers must cope constantly. This issue became the central theme of our study.

As a result of this "reorientation," we added a National Cancer Institute program (Chemotherapy) and SIDS, since both had histories of political system involvement. We dropped Hypertension in favor of two other programs from the National Heart and Lung Institute (Sickle Cell and Artificial Heart); we retained the Genetics Programs as part of the sample.

A few months after the research began at NIH, Bell's agreement to take part led us to a project selection process once again. During this period, we discovered the manner in which the political system seemed continually to exert its influence on research programs at NIH—but pushing for more applied research and for development programs targeted at publicly-espoused health needs. This situation created serious conflict largely because developing knowledge into a service or product often required knowledge and skills outside of NIH's primary expertise in basic biomedical research. Many individuals at NIH believed programs of this type were beyond their normal capabilities and as they were usually expensive efforts, unnecessarily drained increasingly limited resources. The possible accuracy of the latter perception was confirmed in discussions with Bell executives: an applied program could easily cost ten times more than a basic program and a full-scale development effort, 100 times as much.

We used the knowledge gained at NIH to select among the programs presented for study by Bell. Bell Labs had extensive experience in all phases of the R & D process, from basic research through system development, and we anticipated some useful insights and ways of thinking about NIH's involvement with applied and development programs.

Although we interviewed a number of individuals involved in basic research, we concentrated our efforts on product development programs, tracing their history from the "basic" stage to the stage at which we found them. We did not attempt to have "successful" and "unsuccessful" programs represented in the sample. The programs we studied could all be considered "technically" successful, although not all of them had achieved "market" success. We chose two from the Transmission Area: D4 Channel Bank, an evolutionary product that was continually being modified and upgraded; and Millimeter Waveguide, a revolutionary new product that was technically feasible but apparently not destined to make it to the marketplace. From the Customer Services Area we picked the High Capacity Mobile Telephone Program because it was most involved with the political system. And finally, from the Switching Area, we picked the Number 4 Electronic Switching System Program which was the largest program ever undertaken by Bell and which was crossing many of the organization's traditional areas of authority.

In summary, we studied the following nine research programs at NIH and Bell:

NIH: Artificial Heart Research; Cancer Chemotherapy Research; Sudden Infant Death Syndrome Research; Sickle Cell Anemia Research, and Genetics Research.

Bell: No. 4 Electronic Switching System; High Capacity Mobile Telephone; Millimeter Waveguide Transmission; and D-4 Digital Channel Bank.

The "program" was selected as the unit of analysis because it seemed to be the merging point for the management of both scientific and social/political issues. The data collection methods reflected our theoretical concerns, we believe. Because we were concerned with exploring the historical origins of each program and its evolution, our data is essentially retrospective. We have sought in-depth understanding, reliability, and accuracy through both interviews with relevant actors and analysis of documentary evidence of decisions, events, and the paths to them. Because of our limited resources and also because of the desire to explore each case or research program in as much depth as possible, we have limited ourselves to a small number of cases. This number, nine in all, reflects not only our methodology, but also, of course, our financial and time resources. We also have sought not to arrive at a representative sample of cases, but rather to map into our sample the important dimensions and issues which we early perceived in the research. The program sample offered a rich array of experiences for in-depth study. The inquiry was focused on searching for patterns in the management process that could account for varied results. The historical record was considered with emphasis on the senior or strategic management process. This led to examination not only of the internal and technical aspects of each program, but also of their important, multiple external relations. The output of the study is a model that can, hopefully, be useful in ordering the complexities and issues of large research program management.

Summary

Undoubtedly R & D management always has been a challenging job. In the days when it may have been possible to buffer or even wall-off research activity, the challenge was primarily technically-oriented. The organizational systems and processes used or encouraged by a firm contributed to technical excellence. These organizational arrangements have continued to matter because there is a "logic" to research and development activities which, if followed, can produce desired results. Bell Telephone Laboratories has developed an outstanding track record in understanding and managing this logic which produces knowledge and moves it through a series of stages into development and application. We think a lot can be learned from Bell about this process.

But today organization matters for yet another reason. The reality of socio-political intrusion into technical operations is emerging as a problem demanding the attention of managers. Administrators of NIH understand this reality from living with it on a daily basis. Many times in the course of the study we heard comments like, "You can't understand research management here at NIH unless you understand our relationship with 'downtown' " (Congress, HEW, etc.).

A number of the NIH programs provide examples of problems associated with active political involvement in research programs that private sector companies may face to a larger degree in the future. We think AT&T and other firms can learn from NIH's experience.

The effectiveness problems faced by NIH, in particular, were related most directly to problems at the interfaces between the research process itself and the political environment whose financial support was needed, and the social and medical groups which would be using and benefiting from the research conducted. The effectiveness of research programs (as distinct from individual scientists) when investigated, repeatedly led us toward examining the link between the achievement of the scientific task and of successful social relevance. Social relevance was more than a pleasant abstraction; it was a governing condition because the wider environment provided the resources necessary for the continuous survival of the scientific enterprise. Managing that interface by insuring sufficient freedom of movement for achieving scientific goals while producing something of larger value could be said to be NIH's most significant problem.

The problem can be further described as one of metamanagement: the administrative task is to channel and direct a complex social process, involving scientific, political, and social groups with differences of values and goals, in such a way as to preserve an environment that allows science to pursue knowledge development efficiently, but also ensures the development of knowledge relevant to pressing social needs. R & D managers must pay attention to each of these areas or "spaces."

R & D programs can be thought of as existing in three spaces:

Technological space encompasses the physical phenomena on which the R & D organization focuses its technical task effort. For the organizations in our research, it was the discovery and development of knowledge about the human body and communication phenomena.

R & D Management: Old Views and New Realities

Political space is that complex network of relationships between the organization and its environment. Externally, it is the web of the social, economic, and political elements of the environment with which the organization must contend.

Organizational space is the set of internal relationships, administrative systems, and processes of the organization which are utilized to manage and adapt the organization.

Organizations have explicit or implicit strategies, or logics of action, that guide their approach to these spaces. There is a *technical logic* employed in the research process, a *political logic* which constitutes the way in which relations are maintained vis-a-vis the political environment, and an *organizational logic* which gives coherence and continuity to the organization. Often logics which are primarily adaptive to either technological or political space dominate organizational space, with mixed results. R & D organizations need to identify, understand, and ensure the appropriateness of their strategies in relation to these spaces.

The framework we used for analyzing the research process considered each of these spaces and their interrelationships as shown in Figure 2. We used this model to identify and describe patterns in the logics which may have influenced the success or failure of the programs in the study.

We have raised the spectre of increased political intrusion into research and development. Political and technical processes each constitute an important reality for R & D organizations which the next two chapters explore in detail.

Chapter 2 discusses the political context of R & D and describes two programs that existed in volatile political environments.

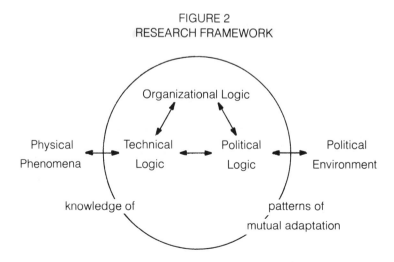

FIGURE 2
RESEARCH FRAMEWORK

Chapter 2

The Political Context of Research and Development

Introduction

As Figure 3 depicts, the political environment focused upon in this chapter is one of the three critical elements in our model

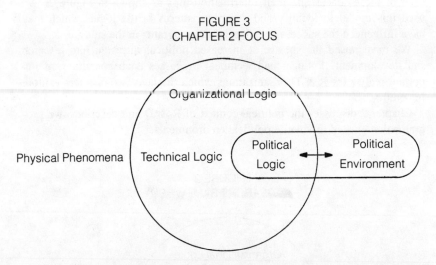

FIGURE 3
CHAPTER 2 FOCUS

The political context of R & D is a network of many inter-related and, often, competing groups. For the programs in our research, it included the government (both elective and administrative elements), organized special-interest groups, including those "representing" the scientific community, contiguous systems of NIH, including other programs and institutes, and the various components of the Bell System.

When the socio-political environment was relatively stable, research and development organizations could be successful by attending to their technical task simply because there were few, if any, opposing viewpoints or priorities. This was particularly true for NIH which historically functioned in a beneficient environment. Congress provided almost unlimited funding; there was little debate about public funding of science or the expected focus or output. Science was self-coordinating, independent, and free to set its own priorities.

The environment changed, however. Vocal special-interest groups rose up loudly to challenge the autonomy of the scientific community and the distribution of resources within biomedical research. All such groups cannot be treated as mere nuisances. As biomedical science and technology increasingly affects society, the public certainly has a right to express its concerns.

Why does an R & D manager in private industry need to pay attention to the political environment? Because technological development no longer is perceived as totally positive contributions unmitigated by offsetting costs. There is no question that some fruits of technology have had a significant beneficial effect on society, but some have not. The public becomes angry when it sees the results of thalidomides and hears of human disability and destruction caused by the presence of industrial wastes like kepone and mercury in water supplies. Three Mile Island has become a symbol for people who believe we no longer are in control of our technology. The demonstrations and fights that erupted at a recent conference on nuclear power in Nagasaki attest to peoples' memory for the danger of radiation and their commitment in opposition to nuclear power.

Not all opposition can be considered bad. Who is to say that the auto industry would not still be concerned with tailfins and other cosmetic features had the public not intruded? Only by pressure did society get economy, safety, and pollution control. Some technocrats may wish these "voices" could be silenced and the socio-political intrusion that they see as having stopped innovation be banished. But the voices will continue to speak and political intrusion will not go away. Industry cannot turn its back on the voices and head for the nearest exist; it cannot ignore its interdependency with the political environment. This relationship must be managed. It has become a fact of life that managing R & D includes managing the interface with socio-political processes.

The formal education and work experience of most R & D managers equip them primarily to understand and manage the technical side of R & D activities. Their background and orientation may not help them in dealing with non-technical considerations. A cognitive map that is useful for negotiating a territory characterized by natural laws and predetermined causal relationships needs to be exchanged for another when the territory changes. A map built on beliefs that science is the apex of the hierarchy of social endeavors and should be self-organizing and autonomous may not fit a territory of social relationships among groups with different values, beliefs, and expectations.

Conflict between extreme values such as "pure science" or "pure business" or "pure social responsibility" becomes more pronounced as resources shrink, as politicians and business executives depend on science and technology to provide the products for continued prosperity and well being, and as researchers look to their firms and government for financing. Members of competing interest groups have been socialized and educated in their groups' assumptions and reasoning and have come, through experience, to believe in the accuracy of the maps based on this logic.

Problems can develop, however, when one group attempts to extrapolate directly from its logic and experience to affairs in another's domain. The temptation for one such "culture" to tell another one how to conduct its business can be irresistible—and possibly dysfunctional. The Sickle Cell Anemia program provides an example of what can happen when researchers ignore the political system; it, in turn, overwhelms the research process, forcing its logic and priorities on a program and dominating it.

A Dominated Program: Sickle Cell Anemia

Sickle cell *anemia* is an hereditary red blood cell disorder that occurs when genes for sickle cell hemoglobin are inherited from both parents. Hemoglobin is the oxygen-carrying component of blood.

The red blood cells develop an elongated, sickle shape rather than their normal spherical shape. The affected blood cells have a life expectancy of 15–30 days compared with the normal 120 days. This puts a strain on the blood producing process and leads to a condition of anemia. The distorted sickle cells also "jam-up" in blood vessels, blocking the blood supply to parts of the body and depriving tissue of oxygen. Most frequently the heart, lungs, kidney, spleen, hips, and brain are damaged. The disorder is characterized by the presence of chronic anemia, jaundice, recurrent bouts of pain called "crises," occasional growth retardation, and increased susceptibility to certain infections. Sickle cell anemia kills many victims before age 20, and few survive to normal life expectancy.

Sickle cell *trait* is a healthy state in which an individual carries genes for both sickle hemoglobin and normal hemoglobin. Rarely are there problems associated with the *trait*, although some individuals may experience some ill effects under conditions of extreme physical stress or lack of oxygen. People usually are unaware of the presence of sickle cell trait unless they are tested for it.

Sickle cell disease is not limited to blacks of African descent, but is found throughout the Mediterranean Basin area. One theory is that it is a natural defense—an adaptation against malaria. It is, however, most prevalent in Central Africa and, therefore, commonly associated with blacks.

In 1971, it was estimated that two million black Americans were carriers of the trait, that between 25,000 and 50,000 individuals were afflicted with the disease, and that approximately 1,000 black infants were born each year with sickle cell anemia.

The Political Context of Research and Development

The molecular configuration of the disease was understood in 1971. At that, more was known about sickle cell anemia than any of the other approximately 150 hereditary blood diseases—yet nothing was known about how to intervene in the process to reverse it, prevent the condition, or successfully treat it. Existing methods of treatment were palliative, aimed at lessening the pain associated with the disease. Although some progress had been made in understanding the disorder, further research, particularly on the sickling process, was necessary.

The Stimulus for a Program

On February 18, 1971, President Richard Nixon delivered a health message to Congress in which he highlighted sickle cell anemia as an important health problem requiring more attention:

> It is a sad and shameful fact that the causes of this disease have been largely neglected throughout our history. We cannot rewrite this record of neglect but we can reverse it.[1]

To reverse the situation, President Nixon authorized a budget increase for research and treatment of sickle cell anemia, from $1 million to $6 million. The Department of Health, Education and Welfare (HEW) and the National Heart and Lung Institute (NHLI) began implementing a Sickle Cell Disease program.

The program constantly was surrounded by controversy. Some people believed it was established only to get votes for Nixon's 1972 re-election bid and that Congress also jumped on the election bandwagon. Others contended that, in spite of possible political overtones, more was known about sickle cell anemia than other blood diseases and that it made sense to focus more heavily on it. In any event, it did come at a time when blacks were asserting their needs and demands. But until this time, apparently few people considered it a major health problem and there were no national special-interest organizations, such as existed for cystic fibrosis or muscular dystrophy, to call attention to it.

How did President Nixon come to emphasize sickle cell anemia in his health message? One account shows that it may have been pretty much the result of circumstance.[2]

In 1971, a black staff member of HEW received a letter asking for help from a woman whose child had sickle cell anemia. The staff member, after

---
[1] "Research, Treatment and Prevention of Sickle Cell Anemia," Hearing before the Sub-Committee on Public Health and Environment, November 12, 1971, U.S. House of Representatives Report No. 92-57, p. 60.
[2] "Sickle Cell Anemia: The Route from Obscurity to Prominence," Barbara Culliton, *Science*, October 13, 1972.

interviewing people at NIH, wrote a report on the disease and sent it to the individual working on the President's health message. This person's father, president of a television station, was shown a copy of the report. After conferences with black leaders in the community, the TV station ran a series of editorials and documentary programs on "the forgotten disease." On one program a professor of pediatrics from a black university talked about his idea of starting a sickle cell center. A goal of $25,000 was established for this cause. Response was overwhelming and $40,000 eventually was collected.

Whether response to this initiative influenced HEW and President Nixon, or whether an emphasis on sickle cell anemia already had been established is speculation; but prior to the health message, publicity was given to sickle cell anemia and public reaction was enthusiastic. The tempo of the times made it a good political bet.

Response to the President's Initiative

At the time of the President's health message, the disease was not being totally neglected. NIH was spending approximately $1 million annually on sickle cell research. Basic research was being conducted at the National Institute of Arthritis and Metabolic Diseases which also was supporting other studies on the disease totalling more than $400,000. The Blood Resources Branch of the National Heart and Lung Institute had awarded $500,000 in grants supporting the first phase of feasibility testing for a therapy for sickle cell "crises." The National Institute of General Medical Sciences and the National Institute of Child Health and Human Development also were supporting some limited scale projects. Sickle cell anemia wasn't exactly the "forgotten disease," but it also was not receiving excessive attention from the biomedical research community; nor was it a disease that the public and the black community, in particular, knew much about.

Although some sickle cell research was going on, there was no organized programmatic effort of the magnitude envisioned by the President. It was obvious to many people at NIH that the underlying causes of the disease were not going to be known for a long time and that program emphasis would be on testing, educating the public, and counseling. NIH personnel with a strong basic-research orientation were disturbed by the political nature of the situation and by the fact that the program would involve educational and other service activities. Most of the institutes involved in sickle cell research were unwilling to assume responsibility for this type of program.

The deputy director of the National Heart and Lung Institute seemed to be the only person at NIH willing to take on the Sickle Cell Program. He had some knowledge of sickle cell disease and an interest in it. His attitude was:

> Here is a problem. If there is something that can be done about it, we should give it an honest try.
>
> The problem was not going to be solved scientifically for a long time. We had

to see if we couldn't solve some short-term problems like adequate testing procedures, educating the public and the medical doctors, and counseling.

Apparently his attitude and knowledge impressed the people at HEW: the National Heart and Lung Institute was named lead agency in the joint federal program which included the Health Services and Mental Health Administration (HSMHA), and he was appointed program coordinator. This appointment was part-time and temporary until a replacement could be found. He remained the full-time deputy director of NHLI.

The Office of the Secretary of HEW structured the program so that the HSMHA would carry out the service aspects and NIH, the research aspects of the program. In addition to its responsibility for research, NHLI also assumed responsibility for coordinating the entire program.

The relationship between the Sickle Cell program and HEW was closer than for most programs. The secretary of HEW appointed a program advisory committee which reported directly to him. The committee was to develop objectives and plans to carry out the President's initiative and to keep the secretary, and the director of NIH advised on program implementation. The committee had eleven members: six with scientific or medical backgrounds and five with backgrounds in social work, business, or public-interest groups. Blacks were strongly represented.

The committee dealt with program objectives and resource allocation issues at its first two meetings. Four objectives were suggested by NIH staff. Two paralleled ongoing activities at NIH: to overcome fundamental biological problems through basic research and to continue development of improved therapies for people in "crisis." The remaining two were to increase public knowledge of the disease through community organizations and to ascertain the feasibility of large-scale screening and counseling programs. The committee modified the objectives primarily by emphasizing the *initiation* of education (community and professional), screening, and counseling programs; but priorities for implementation could not be agreed upon. Funds were divided equally between the two obvious thrusts of the objectives: research and community service.

Further Responses

The sudden publicity about sickle cell anemia generated great interest and some myths. Few people in the black community knew about sickle cell anemia and they became quite interested in learning about it. The program coordinator found himself talking to groups such as the Black Caucus and the Black Football Players' Association and addressing community gatherings in various parts of the country. Local organizations even began sponsoring fund-raising activities.

In the fall of 1971, Congress entered the picture. It held its first hearings on this disease, and introduced legislation which contributed to the ultimate passage of the Sickle Cell Anemia Control Act of 1972. In all, twelve different bills were proposed. The Sickle Cell Anemia Control Act legislated a program of diagnosis, control, treatment, and research and authorized appropriations for the following three years of $25 million, $40 million, and $50 million, respectively. Actual expenditures on the program during those years approximated $16 million, however.

This legislation advocated confidentiality, voluntary screening, voluntary counseling, public education, as well as screening, counseling, and treatment by the military services and Veteran's Administration.

Why the emphasis on confidentiality and voluntary programs? Because of the stigma attached to the disease and the potential for social problems. Demonstration projects in Memphis had shown that if the problem was not presented properly, a stigma immediately appeared and individuals became reluctant to be tested. Even people with sickle cell trait were stigmatized by the language itself. They were carriers or "Typhoid Marys."

Warnings aimed at avoiding social problems came too late: administrative mechanisms to implement screening had been put in place and the stage was set for a monstrous *faux pas*. HEW had designated the Health Services and Mental Health Administration as the organization responsible for the community-service portion of the program. HSMHA, searching for an administrative home for it, chose the National Center for Family Planning Services, probably because the director was black. Implementation by a family-planning agency of a screening program to *identify carriers* conveyed an image of guinea pigs in a testing program and suggested, to some people, impending genocide. This latter viewpoint was reinforced by a paternalistic stance of Congress, and remarks like that of one Congressman who believed the best solution was to prevent carriers of the defective genes from mating and producing offspring. He advocated that individual states require this test prior to marriage.

Testing and identifying people with the trait became a serious social problem. Because it was not known what degree of risk was associated with carrying the trait, blacks became victims of discrimination in buying insurance and in applying for certain jobs where stress or lack of oxygen (airplane pilots, for example) might be a factor. The idea spread that blacks had to be screened for sickle cell trait before hiring.

Is it any wonder the voluntary aspects or the good intentions of the program were suspect? The black community's initial reaction to Nixon's initiative had been: Why did the government wait so long to do something about sickle cell anemia? Let's get going! The identification program, job discrimination, and a fear of genocide combined to reverse this attitude to one of almost total resistance.

Results

In April 1972, a permanent coordinator for the program was appointed. He was a black, a hematologist with considerable clinical experience who had served as a member of the advisory committee. Although he originally intended to occupy the post for two years, he remained for three until the expiration of the Sickle Cell Control Act passed by Congress. He then "returned to science rather than keep pushing paper."

Upon expiry of the Act in 1975, Congress began debate on the program's replacement, if any. The fervor of earlier years apparently had subsided and the thrust[3] of bills introduced in the Senate and House of Representatives was to subsume the sickle cell effort under a program addressing other genetic diseases such as Cooley's anemia and Tay-Sachs disease. This moderate approach was a distinct contrast to the previous bills focusing on sickle cell disease as a discrete entity, and even fostering a dramatic suggestion for a separate National Institute for Sickle Cell Anemia. Eventually, Congress passed an act combining research and community service activities on a number of genetic diseases. Critics expressed concern that funding for sickle cell research would be diluted as it was forced into competition with claims from other disease research.

What was accomplished in the brief life span of the Sickle Cell Anemia Program? Medical and community educational programs were established, testing procedures were developed, and research was stimulated. But an awful lot of work went into overcoming the negative image of the program. Misinformation had to be erased, problems cooled down, and the confidence and support of the black community regained.

In 1976, the solution to sickle cell anemia still was a distant goal. Although some gains undoubtedly were made, they came with a fair amount of anguish and social cost; and the revamped program appeared to assume an orientation similar to the effort prior to political involvement. In 1976 the director of the National Heart and Lung Institute said that the program would continue, but in a more research-oriented direction:

> We're just moving toward more science in sickle cell anemia. I think we've convinced the legislators we're not about to cure it right away and it still requires a basic focus. The educational effort remains but will be more low-keyed and the community service aspects are being assimilated by neighborhood clinics which are multi-categorical and not just devoted to sickle cell anemia.

Commentary on the Sickle Cell Anemia Program

The sickle cell case illustrates the potential effect of political "intrusion" into R & D. A cause is temporarily championed, public expectations are raised beyond the delivery capability, tension increases, some people are hurt, and interest wanes. The research organization is left with bitter memories, to take the blame and pick up the pieces of a rush job; and the problem is not solved.

---
[3]The thrust of these proposals can be seen in H.R., 7988, October 21, 1975. U.S. House of Representatives.

The Sickle Cell Program existed in the most volatile political environment of any of the cases we studied. It was a period of strong black consciousness and a struggle for human rights. Responding to the needs of the black community might have resulted in electoral advantage and thus the decision to establish the program possibly was influenced by political considerations. Although more was known about sickle cell as a genetic blood disease than any other such disease, it is not at all clear that the state of knowledge justified the special programmatic backing that emerged.

It would have been difficult for the program to have achieved much in the way of research effectiveness with the limited time available to it. Any major scientific breakthrough presumably would have been serendipitous. This was a major constraint on the program's capacity to grow and survive. Initial achievements in the service areas could not be used to improve relationships because of the mismatch between public expectations and scientific reality.

In sickle cell, the political system initiated the issue and forced it into national prominence. It is possible that the politicians' interests were paramount and NIH's research capability became a political tool. NIH, as a government research institute, is more "open" to its political environment than Bell Labs. A certain degree of protection is essential for the development of effective research programs, but there is an important tradeoff. Too much insulation can cut off the research effort from important environmental stimulation and a significant source of feedback. Too little insulation can result in a failure to maintain a program's integrity. "Protection" can take many forms. Structural and legal defenses spring immediately to mind, but are not necessarily possible in a place like NIH. Proactive engagement of the political system in helping shape a program is another form.

Sickle cell is an example where NIH remained generally passive in an attempt to avoid service-activity involvement. The strategy of avoidance did not work. The intention of the political system was not to be denied and it probably was fortunate for the integrity of the larger NIH system that one individual was willing to get the program going. The political system exerted considerable influence in the choice of the research task and the administrative structure of the program. The organizational structure of the program was dictated by the Executive Branch and reflected necessary social and scientific dimensions in a deliberate division of task responsibility. This differentiated structure, while, perhaps, a necessary condition for success, was not sufficient. A synthesis of the two domains was not achieved, and potential social problems either were not considered and/or understood, or were ignored in the rush to implement the program.

The experience of this program should constitute an important lesson for others at NIH in similar circumstances: If you do not pay attention to the political system and you try to ignore its initiatives, your technical process runs the risk of being overwhelmed.

Shifts in the Relationship with the Political Environment

Achieving a stable, balanced relationship with the socio-political system can be a major managerial job. Once attained, it may be dangerous to assume that the job is complete. The relationship must be maintained in an environment that does not stand still.

At Bell we studied a program in which a once stable relationship came apart and we learned about the struggle to readapt. The High Capacity Mobile Telephone System (HCMTS) bogged down in controversy and was delayed. The principal actors in this drama were AT&T and other providers of mobile service, customers, equipment manufacturers, and the federal government.

Destabilization and Readaptation: High Capacity Mobile Telephone System

Although mobile communication service had been available since 1946 and considerable knowledge had accumulated about radio propagation, the High Capacity Mobile Telephone System (HCMTS) was a totally new concept made possible by 1960's research into voice propagation at very high frequencies. Mobile service was defined as radio communication between a moving vehicle and a fixed station through the existing telephone network.

Mobile service could be purchased from two sources:

1. *Wireline companies*. These were the "traditional" telephone companies (Bell, General Telephone, and Continental) that owned existing telephone lines.

2. *Radio Common Carriers (RCCs)*. These independent operators owned radio transmitting and receiving equipment connected to the telephone lines. They offered mobile telephone and paging service to customers in a local area. The largest, in Chicago, had approximately 600 mobile and 6,000 paging customers. A typical RCC might service 200–500 customers. The common interests of the RCCs were represented by an industry association, the National Association of Radio Systems (NARS).

While the RCCs and wireline companies were competitors, they did not compete within the same radio band. Radio space, a scarce resource, was controlled by the Federal Communications Commission (FCC) which allocated each entity (wireline companies and RCCs) the same number of channels in two separate parts of the radio spectrum. Thus, while Bell competed with RCCs for customers, it did not compete for space in the radio spectrum.

Mobile systems typically used broadcast techniques similar to radio or television stations. High towers and high transmitting power provided large service areas with approximately 20-mile radii. While it was an efficient way of reaching a person when only a relatively few radio channels were needed, the technique precluded reusing the same channels within a 75–100-mile radius. To avoid interference, a radio channel was not reused within a distance four to five times the radius of the service area. This was a serious constraint in large metropolitan areas.

Existing systems increased capacity either by adding frequencies or by adding customers to each channel. The number of frequencies was limited, and adding customers to existing channels ultimately resulted in poorer service.

Neither the customers nor the providers of mobile telephone service considered existing service adequate. For years, the wireline companies and the RCCs prodded the FCC to rectify the deficiency by making more spectrum space available, and in 1968, the FCC began listening to recommendations. Ultimately, in May 1974, the FCC granted additional radio space for the development of a high capacity, cellular-type system.

The cellular concept, originated by Bell, accommodated growth by permitting channel assignment and reuse on the basis of population density. A geographic area was divided into cells and low power transmitters, which were only strong enough to cover that area, were installed. By reducing the radius of radio coverage from 20 miles to between one and eight miles, the expansion capacity of the system was greatly increased.

Controversy began when the FCC issued its report in May 1974. Prior to this report there had been no direct competition for radio space between the wireline companies and the radio common carriers (RCCs). However, the additional spectrum space allocated by the FCC was for operating cellular systems, and only one cellular system was permitted in a given geographical area. Now there was direct competition between the telephone companies and the RCCs.

The FCC established five important conditions:

1. Only wireline companies could install high capacity, cellular systems. RCCs were excluded because high capacity systems required extensive capital investment beyond the capabilities of the RCCs.

2. Telephone companies were not permitted to manufacture radio transmitters, receivers, or mobile units.

3. Telephone companies were required to establish separate organizations for mobile service to avoid cross-subsidies. These new organizations were to have separate accounting records so that contribution to the parent organization could be readily determined.

4. Telephone companies could not own the mobile receiving units. Cus-

tomers would have to purchase them ($1,200-$1,500) or lease them from a manufacturer.

5. On the basis of trial system performance evaluations, a set of national standards would be determined by January 1979.

The FCC also created entities called Specialized Mobile Radio Systems (SMRs) which were to operate as additional RCCs. The mobile equipment manufacturers were eligible to become SMRs. The FCC allocated them separate spectrum space and proposed they be allowed to operate unregulated by individual states while telephone companies and RCCs continued to be regulated by the states in which they operated.

Needless to say, this situation created a storm. The RCCs were up in arms. They always had had equality in spectrum space but were now, in effect, being eliminated. The National Association of Radio Systems (NARS) argued that:

1. The creation of SMRs was illegal.

2. Restricting cellular system development to the wireline companies effectively eliminated the RCCs from competition. NARS argued that if it were true that only telephone companies had the capital to develop and install these systems, then the condition would arise naturally, and thus, there was no requirement to make it a regulation. NARS believed cellular systems were unfair to their interests and free competition and should be prohibited.

Bell also contested the conditions in FCC administrative hearings. It argued the following points:

1. There was no need to have a separate entity operate the system in order to prevent cross-subsidization.

2. More spectrum should be allocated so that systems could operate more economically and efficiently.

3. The telephone companies should be allowed to buy mobile equipment and lease it to customers. It was too expensive for any entrepreneur or even an equipment manufacturer to build, stock, and lease the mobile units. This could limit availability of mobile units and retard growth of the system. Apparently, manufacturers agreed, favoring the stability and predictability afforded by large purchase orders from the Bell System.

In March 1975, the FCC modified its position. It took cognizance of NARS' argument and lifted the restriction limiting cellular systems to wireline companies, and permitted telephone companies to purchase mobile units and lease them to customers.

The RCCs were not satisfied by this outcome and NARS hired a former Supreme Court Justice to contest the FCC ruling in court. In January 1976, the court found in favor of the FCC rulings. Cellular systems could be utilized, open competition for developmental systems and the existence of SMRs were declared valid. NARS appealed this ruling to the Supreme Court, which decided not to hear the petition.

The controversy also included arguments by NARS and equipment manufacturers against Bell's proposed developmental system. To achieve economic growth, Bell developed a start-up configuration for Chicago that differed from the mature cellular configuration as initially proposed. This was a more economical mode of implementation potentially saving $75 million in start-up costs. As the number of customers grew from the initially anticipated 2,000, this configuration could accommodate up to 20,000 customers before cell-splitting and the mature configuration would be necessary. Other than a different antenna-type, all other proposed features were the same for the start-up and mature configuration.

NARS and the large RCC in Chicago attacked Bell's proposed developmental system on this modification in order to forestall Bell's introduction of the system in Chicago.

Equipment manufacturers also argued against this system because they felt that Bell's method of utilizing the spectrum space "sterilized" the spectrum and made it impossible for others to use their portion differently. They were adamant that the separate operating company be established immediately to avoid cross-subsidies during development.

Prior to the Chicago system challenge, the HCMTS program was basically open to the FCC staff. Program personnel from Bell Labs met frequently with the FCC staff who often attended in-house technical reviews. These challenges to HCMTS altered Bell Labs' traditional interaction with the FCC. A somewhat adversarial, arm's-length relationship developed and AT&T's Legal and Regulatory Divisions became the contact with the FCC.

The Nature of the Political Environment

We see in this situation the nature of the political environment. It is a socially constructed reality shaped through the competition of multiple value systems of which a subset eventually becomes dominant and binding on the actors in the system. In negotiating the parameters of this environment, the actors engage in a process of value-advocacy (Vickers, 1973). The values and logic of each organization are contested by others in a continuing process until resolution by either the parties themselves or by some external authority. The domain, rights, and responsibilities of each are then defined, outlining the boundaries and interrelationships that characterize this political eco-system. This environment is one where organizations appear and disappear not only through direct competition, but also by decree.

Prior to May 1970, socio-political relationships appeared to be in balance. The wireline companies and the RCCs evenly shared the scarce resources of radio spectrum space with no competition for space and minimal competition for customers, since there was a waiting list for service. The equipment manufacturers' sale of hardware to both the wirelines and RCCs was their

The Political Context of Research and Development

only involvement in the mobile service market. Mounting customer demand for mobile service ultimately upset the system's balance. Changes in technology and spectrum allocation were necessary to satisfy demand, but the industry structure held in place by then-existing FCC regulations was dramatically changed.

The proposal to expand customer service capability and to introduce new technology spawned complications, upsetting the old ecological balance. Members of this political system were left competing to protect their interests by attempting to influence the architect and architecture of the new eco-system. The FCC had the power to dissolve existing rules and interrelationships, define and enforce new ones, and even change the mix of actors in this eco-system. Initially, only the wireline companies were going to be allowed to develop new technology; direct competition for spectrum space was introduced; and new entities (SMRs) were decreed into existence and provided with resources and conditions for competition (separate spectrum space and freedom from state regulation).

The RCCs could not be competitive individually with Bell in developing cellular systems in view of the capital investment required, and saw themselves being eliminated by the new direct-competition conditions. Through administrative and judicial appeals, they claimed the value of efficiency from multiple competitors to be in the best interest of the consumer, and attacked the start-up of the Chicago test system on what appeared to be a technicality. Bell, on the other hand, claimed the value of efficiency made possible by a large organization to be in the consumer's interest. The SMRs argued that the wireline companies' spectrum share was too large. Presumably, any residue from a reduction would accrue to them. They also argued for the efficiency of certainty in equipment production and orders, made possible if the wireline companies could own and lease mobile radio equipment.

The social problems and challenges presented by the socio-political environment differ dramatically from those presented by the natural phenomena of the physical world where R & D managers may be more familiar and comfortable. But the environment is a reality to which management must adapt proactively and help to shape rather than simply react and risk unfavorable consequences. A relationship of some sort, either positive or negative, between the R & D organization and constituents of its political environment is certain to crystallize.

Patterns of Mutual Adaptation

The interdependency between research organizations and other organizations such as Congress or special-interest groups may be more significant than research administrators would like to admit. Research has become institutionalized, taking place in complex organizations which must continually interact with these groups. The exchange is generally a return of new knowledge or of products which contribute to the goals of a larger system for investment of resources—money, people, space, equipment, and so forth. The goals of the larger system may be significant constraints on research programs. Research program managers must know and understand these objectives and factor them into their planning and operations. Often the locus of control or power over a research program lies outside the program.

An easy, but possibly short-sighted, response to the reality of political intrusion is to pretend it doesn't exist. This situation is similar to a condition Vickers (1968) termed the "end of freefall"—a process similar to a man falling from the top of the Empire State Building who was heard to say as he passed the second floor, "Well, I'm alright so far!"

Weinberg's (1967) analysis of the growth of science as an institution provides a useful analogy. For science, the end of free fall may have begun as it made the transition from "little science" to "big science."[4] Its requirements for resources were growing as was its dependence on other social institutions for support. Science expanded, became highly differentiated, and drifted away from the social realities with which the political system is concerned. A value barrier developed, tending to seal science off from its supporting environment. An attitude placing science at the apex of social activities emerged.

Weinberg observed that government has two types of choices regarding resource allocation. It may choose to support science in relation to other competitors for funds such as universities, defense, housing, and welfare. Its second choice is the fields within science to support—biology, physics, oceanography. The evaluative criteria should ask "why" support these undertakings. This is a dilemma recognized by administrators from all institutions—competition in an arena of multi-valued choices.[5] Is an extra pound of science worth the same as a pound of housing? Or what about the relative worth of biology and physics? It is the policy problem faced by everyone attempting to achieve multiple goals with limited resources.

Science, in its turn, should be responsible for choices within disciplines and for making the evaluation about how resources should be spent. The separation of "why" and "how" is important. If a single system can determine both why and how, the danger of its sealing itself off from reality increases. In Weinberg's opinion, only external criteria generated outside a given, closed system of logic can ask "why."

Applying Weinberg's idea that the political system has a legitimate con-

---

[4]The terms "little science" and "big science" come from the book, *Little Science, Big Science* by Derek J. deSolla Price, Columbia University Press, 1963.

[5]For an enlightening discussion of the policy-making process that addresses this issue, see *A Strategy of Decision* by D. Braybrooke and C. Lindbloom, *The Free Press*, New York, 1970.

cern in balancing research priorities with social needs, but does not necessarily have the expertise to resolve or even recognize the scientific uncertainties, we saw in the Sickle Cell Anemia Program an instance where the political system apparently exceeded the bounds of its capability. This logic suggests, however, that the scientific community waits at an imaginary dividing line primed to do its part in deciding the "how" and "how fast." Unfortunately, this is not always the case. "Corporate" NIH never assumed responsibility for dealing with the issues in that situation.

The end of free fall has arrived for some organizations and may not be far away for others. For private R & D organizations, this process may have started when some of their fruits displayed detrimental side-effects.

These organizations need to maintain control over the program task structures, technical evaluation systems, internal resource allocation (Weinberg's "how"), and their portfolio of research projects. But this internal focus may not be sufficient for success any longer. Firms must be responsive to legitimate external concerns, while protecting themselves from unreasonable demands.

In this study we observed three basic relationships, or patterns of mutual adaptation, develop between programs and their political environment. Two of these relationships proved detrimental to the technical progress and output of research programs. The espoused benefits did not accrue and considerable amounts of money were spent. The two patterns of mutual adaptation that were unstable and that had negative effects on research programs could be labeled domination and rejection.

Domination: Situations can develop in which the necessary balance between scientific and social priorities is absent and one or more environmental constitutuencies force their priorities on a program and dominate it. For example, the political system may significantly increase the resources for a research program benefiting a particular group in hopes of gaining votes. This increase may take place even though the money cannot be spent effectively by the scientific community on research, as was the case in the Sickle Cell Anemia Program.

Rejection: Rejection is a pattern of adaptation in which a research program and significant constituency groups disengage from each other and become mutually isolated. In the case of the Artificial Heart Program, the unresolved conflict between the program director and the director of NIH about the nature of the program created a breach. This relationship dominated the history of the program and eventually debilitated it. We will discuss this program in Chapter 4, focusing on the impact of the social relationships on its technical task.

We also discovered a pattern of mutual adaptation that contributed positively to the on-going research and enhanced the program's image in its socio-political environment:

Symbiosis: A symbiotic pattern of mutual adaptation exists when a research program and its social constituency groups exist in a mutually beneficial relationship without either the program's, or its constituencies' unilateral domination. Such was the situation with the Sudden Infant Death Program, the scientific community, parent interest groups, Congress, and administration of NIH. We will see how this relationship came about in Chapter 5, and how it protected the program's research focus.

Achieving a symbiotic relationship does not mean the adaptive process has brought the organization to its ultimate resting point and can be forgotten. The organization has become adapted, but it must remain adaptable. In the HCMTS Program we saw a one-time symbiotic relationship become unstable and the participants struggle to re-adapt.

Why should managers pay attention to their political environment? Because it is inevitable that some form of relationship will develop and the danger, from a research program's perspective, is that this relationship may end up misdirecting an appropriate technical logic.

Chapter 3

The Technical Logic of Research and Development

Socio-political intrusion cannot be made the scapegoat for all R & D problems or every program failure. The R & D process itself is more complex and dynamic than is usually recognized. It does not just unfold deterministically with each discrete task reaching final completion and neatly blending into the next stage of development. The subject matter on which research programs concentrate is generally inherently complex, and the job of deciphering the workings of physical phenomena and discovering causal relationships is uncertain and ambiguous. Successful research and development does not occur automatically; instead, it is a result of careful management based on a comprehension of the intricacies of the R & D process and a realistic assessment of potential outcomes.

Successful research managers rely on their knowledge and experience in charting a path through the R & D maze with all its starts, stops, and redirection. Reliance on informed judgement is essential in uncertain environments, but this also means that even the most capable managers may make mistakes. Even without the confounding effect of political intrusion, research programs may go astray if they lack a thorough analysis of constraints such as a realistic appraisal of the type and level of activity that existing knowledge will permit. Diminished performance levels may well result, regardless of whether the reason for the incomplete analysis is ignorance, oversight, or a temporary blindness created by overly optimistic hopes and expectations.

A reasonably discriminating map that is useful in R & D territory can aid judgment and help improve the success rate for programs where political intrusion is not a factor. Where intrusion is a factor, such a map is an essential tool for communicating with elements of the political environment that want to initiate unrealistic programs or to redirect existing research efforts. Policy makers in such a situation need a cogent explanation of the technical and fiscal realities of their proposed endeavors, including the potential risks and reward, as well as a realistic estimate of time to completion. If these interactions are handled well, it should be possible to establish satisfactory cost/

benefit ratios; the research programs should be able to create the operational space they need, and all parties stand to gain from the transaction.

In conducting our research, we looked for a map of the R & D process that would let us order the facts of our sample programs and compare these data. Probably because of the unique nature of the programs and institutions with which we were dealing, we did not find such a tool and thus developed our own. This chapter describes this map which we call the technical logic of R & D and which we will use in the following two chapters to analyze some of the programs in our study. Figure 4 depicts the focus of this chapter.

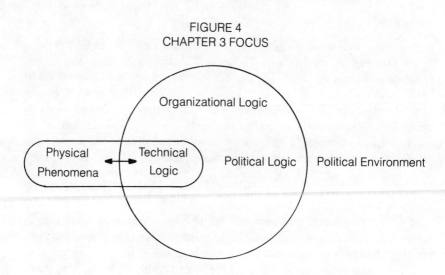

FIGURE 4
CHAPTER 3 FOCUS

The Technical Logic of Research and Development Programs

During our research, we found the standard conceptualization of the R & D process limiting. Descriptive terms such as "basic" or "applied" were difficult to relate to organizational processes. Words can obscure essential elements of a process like research and development. People attach values to symbols and often transmit the values rather than details of the actual process. We found it necessary to look behind these labels to uncover some important characteristics of the research task itself.

Gordon characterized the standard terminology as preconcepts which must be overcome for progress to be made in understanding and improving R & D management. He stated:

The Technical Logic of Research and Development 35

> ... the dichotomy, basic and applied (and such related dichotomies as limited and fundamental, theoretical and action, et cetera) hinders rather than advances our knowledge of the relationship between organizational structure and scientific accomplishment. The distinction between basic and applied research is not directly related to the research process and, therefore, can only be of limited value in research administration. Further, the mystique around the terms "basic" and "applied" appears to have led to patterns of research administration based on misconceptions rather than on knowledge.[1]

Standard R & D stage typologies, although suggestive of a flow of knowledge from "basic research" to "applied research" and into "development," are basically static. These typologies have had multiple objectives: suggesting a work flow, describing the researcher's viewpoint, and conveying the commercial implications of the research.[2] Trying to capture important technical and social complexities in a one-dimensional continuum oversimplifies and obscures some critical organizational processes such as transferring knowledge produced in one part of the organization to other parts, or the organization's response to the concerns of groups in the environment. Mixing social and scientific dimensions obscures the important reality that scientific and social priorities can be mutually exclusive, and that consensus may be difficult to obtain. The result may be conflict between the organization and elements of its eternal environment or, internally, between researchers and management.[3]

Researchers often pose problems that are solvable, while society often wants more intractable problems worked on. An administrator at NIH provided an example of this conflict:

> The logic of genetic disease demands that one thing be wrong in many cases. In sickle cell anemia there is a wrong amino acid; in PKU, a faulty enzyme. There are also multifactorial problems which involve multiple genes or environmental insult. Single gene research is the most fruitful; multifactorial, the most important. Many researchers choose single gene research because they can get answers.

The dilemma is clear. There is a need to mediate conflicting priorities. We believe an approach that separates the technical and socio-political aspects of research and development and then focuses on their integration will help increase understanding of both the R & D process and the way research organizations adapt to their political and technical environments.

Writers on innovation have added a dynamic orientation. Innovation is

---

[1] Gerald Gordon, "Preconceptions and Reconceptions in the Administration of Science," in *Research Program Effectiveness*, Marshall C. Yovits et. al., editors, New York; Gordon and Breach, 1966.

[2] Lowell W. Steele, *Innovation in Big Business*, New York; American Elsevier, 1975.

[3] Richard S. Rosenbloom and F. W. Wolek, *Technology and Information Transfer: A Survey of Practice in Industrial Organizations*, Boston; Division of Research, Harvard Business School, 1970.

seen as a process incorporating a sequence of events from invention through implementation. The innovation literature did not solve the problem of interweaving social and technical issues, but it did relate the R & D process more closely to social processes by emphasizing implementation and product acceptance. This literature recognizes a wider environment possessing some form of approval or veto over products presented to it.

The innovation process begins, in the literature, with the synthesis of knowledge into a product. Knowledge exists that can be fashioned into an "invention." Although an improvement over static R & D typologies, most of these studies were of industrial R & D firms and did not follow the knowledge-generation process back far enough for our purposes. Much of the research at NIH and Bell Labs is at a still earlier phase that we will refer to as *discovery*.

The difference between discovery and invention is subtle but, we believe, important. To invent implies bringing something new into existence; a making as opposed to a finding. To discover is to obtain knowledge of for the first time; discovery implies the pre-existence of what becomes known—a finding rather than a making.

The organizations in our research were heavily committed to R & D and innovation. Research into the structure and processes of natural, physical (biomedical and communication) systems was a critical element of their primary mission. NIH was oriented more toward scientific research, with developmental work representing the smaller part of its activities. AT&T was oriented more toward development, but fundamental research was an important factor in AT&T's continued success. Although the mix of activities in each organization differed, the technical task of both was to generate knowledge of natural phenomena and transfer this knowledge into practice. Few organizations have been as successful in this endeavor as these two have been.

We have defined technical logic as the constellation of assumptions, ideas, and concepts focused on technical achievement. It is the organization's theory for accomplishing its primary technical task. It is the map used by scientists to find their way in technological space. Here, we first describe the development of genetics knowledge and the five basic steps in the technical logic of biomedical research. Next, the comparison and combination with the R & D sequence of AT&T provides the model of technical logic we used in our research. Finally, the Millimeter Waveguide Program illustrates the technical logic in action.

The Development of Genetics Knowledge

Prior to the publication of Gregor Mendel's work in 1865, the common explanation for inheritance of traits from generation to generation was a physical mixing of the parents' blood. Mendel, a monk, demonstrated that certain variations in pea characteristics, such as color or size, could be accounted for by mathematical relationships between their observed characteristics and those of previous generations of peas. Mendel theorized that transference of such traits was controlled by discrete units contributed by each parent. These units, or genes as they became called, did not mix, but somehow interacted to produce the particular pheno-typic characteristics of the next generation. For a long time, genes were merely a concept and beyond the realm of experimental science. It was not known where they were located, what their structure was, or how the transference process worked.

During the 1930s and 1940s, scientists sought to identify the carriers of genetic information. Around 1944, O.T. Avery showed that DNA could be such a carrier. In 1953, the Watson and Crick discovery of DNA structure marked a significant point in the history of genetic research. Researchers had needed to identify a biochemical unit that could perform three functions: reproduction, information storage, and direction for the growth of the human cell. Watson and Crick's work resolved this problem and in so doing turned genetic research from "problem solving" into "puzzle solving."[4] Once the nature and structure of a physical phenomenon is identified and described, it becomes possible to conduct systematically a series of puzzle-solving experiments to learn the function of sub-units, their interrelationships, and their roles in the larger system. In this sense, genetics research became a puzzle-solving science after 1953.

A range of puzzles needed solution. There were broad issues, such as the composition of the genetic code, that set of instructions encoded in DNA which ultimately determines the nature of living matter. There were also specific puzzles for each genetic disease. What combination of faulty components and interactions are expressed as a particular genetic disease?

The cracking of the genetic code permitted more detailed research into specific diseases. Some researchers approached this puzzle as a mathematical problem similar to solving the secret military code, while others experimented with physically altering DNA and observed the consequences in the end product.

The classic code-breaking experiments were conducted in late 1961 at NIH, but it wasn't until 1966 that the code was completely deciphered. It was a complicated puzzle!

Once the genetic code had been cracked, two divergent avenues for future research appeared: a continuation of basic studies of molecular level events and the chemical composition of genetic materials, or a move toward a more applied orientation focusing on human genetic disease. Genetics had the choice of moving from an observational to a manipulative science.

For those researchers interested in studying human genetic diseases, two

---

[4]Thomas S. Kuhn, *The Structure of Scientific Revolutions*, The University of Chicago Press, 1962.

new factors rapidly emerged. No longer were model systems, such as fruit flies, adequate. Human cell tissue was needed, but it was technically and morally impossible to experiment on humans. A special bank of human cell tissue had to be created and maintained so that experiments could be carried out freely without harming living humans. Secondly, the public was becoming much more interested in genetics. The dramatic scientific progress raised both expectations and concerns about the uses of this knowledge.

Clearly, genes had existed and performed their functions for a long time prior to our awareness and knowledge of them. Having described briefly how scientific knowledge of these natural objects changed genetics research over time to a series of more tractable puzzles, we now look at the logic underlying this knowledge development.

Technical Logic of Biomedical Research

There is nothing esoteric about the idea of a logic to the process of scientific discovery. A number of interrelated phases to knowledge generation can be expressed as a theory with reference to a particular field or phenomenon. Theorists writing about the administration of research and development have put forward a type of technical logic. The sequence of basic, applied, and development phases in the research process is a theory or logic of knowledge development. It is a statement of a set of interrelated phases, not randomly interrelated, but patterned such that basic research precedes applied research and so forth.

Genetics research provides a clear example of a technical logic for biomedical science. Its five phases are described in sequence. The work of Watson and Crick constituted the culmination of an *identification and classification* phase in our knowledge of human genetic disease. The sequence of development from Mendel to Watson and Crick comprised an attempt to define specifically the genetic units and to classify the important molecular structures and their roles in genetic development. The discovery of the DNA double helix did not, however, provide information on the natural progression of specific human genetic diseases.

The next step was to isolate the differences between normal and abnormal states of development. A researcher may, for example, wish to know what is different about the genetic and chromosomal structure of individuals with Downs Syndrome compared to those without it. At this stage, no attempt is made to "explain" the syndrome or to discover its basic cause, but rather to see, for example, if certain genetic characteristics occur in the parents, or if the syndrome correlates with the presence or absence of certain enzymes. In a sense the syndrome is still a black box. Questions such as these represent the process of *comparison*, the second stage in biomedical knowledge development. Researchers develop understanding of the relationship between particular genetic problems and the presence of certain genetic materials.

The third phase, which is still largely incomplete in genetics research, is an attempt to understand the *mechanism* by which certain starting conditions and biomedical processes lead to the specific disease characteristics. Researchers must link anomalies in the genetic material with a set of intervening events that leads to a particular genetic disease. For example, the mutation of the DNA structure may inhibit the production of a certain enzyme which inhibits the production of a protein needed by a particular organ, and so forth. It is necessary to understand the causal linkages between structure, process, and final event. The mechanism by which the abnormal event is expressed is unique for each genetic disease or syndrome.

Once the mechanism is understood, science then can move into a phase of *intervention*. Experiments can be conducted with diseased systems to correct the condition or alleviate the consequences. If the disease mechanism is understood, researchers know theoretically where to intervene, when to intervene, and how much to intervene in order to effect desired results. Knowledge of some genetic diseases is now at a stage where intervention can be attempted. The testing of amniotic fluid in a pregnant woman is a mode of intervention which, together with abortion, can prevent the birth of children with Downs Syndrome. Ultimately, it would be technically desirable for intervention to occur at any point in the chain of development to correct genetic problems.

The fifth and final stage of knowledge development is a program of *clinical trials* to validate any mode of intervention. Genetics research, to our knowledge, has not entered this phase, but such clinical trials would be needed to operationalize the modes of intervention designed by medical science.

This technical logic of biomedical research and its application to genetic research are summarized in Table 1.

With the progression of genetic research and knowledge building came changes in the operational aspects of the research task. Comparison of the early and later stages of genetic research reveals a change in four characteristics. First, the necessary resources for the research were relatively inexpensive compared to demands of later stages. Second, researchers had a common interest in, and orientation to, the primary problem of unraveling the genetic mystery. Third, the raw materials for experimentation were "socially free." This does not mean that there was no expense. It costs money to breed fruit flies and keep them alive under controlled conditions. However, we doubt that anyone was morally concerned about the use of fruit flies in experimentation. In this sense, they were a socially free good to be treated in an arbitrary fashion by the scientific community. Finally, the organizational complexity of conducting the task was relatively low and it was necessary only for small teams of like-minded scientists to be involved in any particular experiment. The complexity of the field was contained within a fairly unified and small, social group of scientists sharing a common research viewpoint.

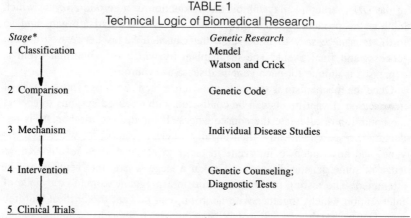

TABLE 1
Technical Logic of Biomedical Research

Stage*	Genetic Research
1 Classification	Mendel
	Watson and Crick
2 Comparison	Genetic Code
3 Mechanism	Individual Disease Studies
4 Intervention	Genetic Counseling; Diagnostic Tests
5 Clinical Trials	

* These stages were suggested by Dr. Brian Kimes of The National Institutes of Health.

In the period after the cracking of the genetic code, these characteristics differed considerably. There were no specialists in human genetics compared to the numerous specialists in molecular biology who were concerned with "basic" genetic processes. The financial resources required to support the research on human genetic disease were considerably larger than those needed to investigate the structure of DNA. Tissue cell banks, genetic counseling, and intervention into human systems are expensive. The raw materials no longer were socially free, in that the public became both morally concerned and involved in the process. Researchers cannot treat human beings in the same cavalier fashion as fruit flies. Finally, the complexity was greater and not contained. The actual nature of human genetic problems is more complex and needs multi-disciplinary physical and social science involvement in order to achieve successful research output.

Any research program organized to meet the pre-genetic code conditions would be faced with considerable adaptation to meet the changed conditions. The Genetics Program at NIH adapted to these changes quite well as we will explain in detail in a later chapter.

The Technical Logic of Research and Development

Comparing the Technical Logic of Biomedical and Communications Research

The accumulation and development of knowledge in the field of genetics research provided an opportunity to explain the five basic stages of biomedical research. In summary these were:

1. *Classification:* This is the process of identifying and classifying natural phenomena.

2. *Comparison:* Normal and abnormal systems are compared in a search for similarities and differences. This is a quantitative stage where differences can be measured, but their cause is unknown.

3. *Mechanism Studies:* This is a qualitative stage of deciphering and understanding the causal mechanism. It is a critical point, since understanding the causal mechanism permits development of intervention technology.

4. *Intervention:* Researchers start developing tools for diagnosing and preventing a disease and begin intervening into diseased systems.

5. *Clinical Trials:* Potential curative agents based on research finds are tested in human systems.

The technical logic used by the Bell System for communications research and development follows the steps shown in Figure 5, which is very similar to the standard typologies found in the literature describing the stages of R & D or the innovation process. Fundamental or "pure" research projects are undertaken in fields considered potentially relevant to Bell's business. These fields include chemistry, metallurgy, psychology, biophysics, and genetics. The output, in forms such as talks and papers, *possibly* may be useful to Bell in a yet-undetermined way in the future. Exploratory development is focused on achieving a fuller understanding of fundamental research output, particularly that which is potentially feasible technically and *probably* useful in the Bell System. The remaining three stages—development, production, and implementation—concentrate on the tasks that the names suggest for specifically selected systems.

FIGURE 5
TECHNICAL LOGIC OF COMMUNICATION SYSTEMS RESEARCH

Comparison of the biomedical and communications research sequences reveals an overlap. The classification, comparison, and mechanism stages parallel fundamental research, and the intervention and clinical trials stages are similar to exploratory development. In developing a more generalized typology it is important to understand the intent and constraints of each stage. Our scheme attempts that goal.

The model we found useful for comparing the research programs in our study has seven stages:

Stage 1. Classification
Stage 2. Comparison
Stage 3. Mechanism Studies

Stages 1–3 represent the process of searching for and understanding the causal mechanism or critical linkage of a physical or biological phenomenon and some event of interest. We will often use the term *discovery* as a shorthand notation for these three stages.

Stage 4. Exploratory Development. The concern at this stage is to identify, understand, explore, and control the important boundary constraints imposed on the new technological development by the larger system of which it will be a part, or into which it will intervene. It focuses on understanding and controlling the interaction between a developmental system and its ultimate physical or biological environment.

Stage 5. Development. This is the process of translating the scientific and technological knowledge previously acquired into specifications for production, including development of production processes to make manufacture possible. A decision to continue into production implies assurance that at least one way of doing the job exists.

Stage 6. Production. The efficient manufacture of the developed system.

Stage 7. Implementation. The process of introducing into general usage the treatment, cure, product, or system.

A critical dimension of research is the understanding of causality. Once the concept of a causal mechanism or process is established, two often competing research logics can be distinguished—sequential and empirical. The *sequential* approach (Figure 6) to research is a step-by-step process designed to discover the causal mechanism.

The *empirical* approach (Figure 7) on the other hand, assumes that disease can be controlled or patients cured without ever totally deciphering the causal process.

The empirical logic has been successful in a number of instances, for example, developing antibiotics and preventing polio. There seem to be situations in which the empirical strategy is fruitful, in such cases as a unifactorial disease or a relatively simple biological system.

We suggest that these two strategies can best be seen as alternative ways to carry out the exploratory development phase of the model. If the mechanism that is the critical causal link between biological phenomena and a disease is well understood, a systematic engineering effort, for example, may be a relevant approach. If the mechanism is not understood, the empiric model can be considered a calculated high-risk strategy to attempt to find a solution without the benefit of a complete discovery phase. Utilization of either approach, however, should be based on knowledge of inherent risks and of the potential drain on resources.

FIGURE 6
SEQUENTIAL RESEARCH LOGIC

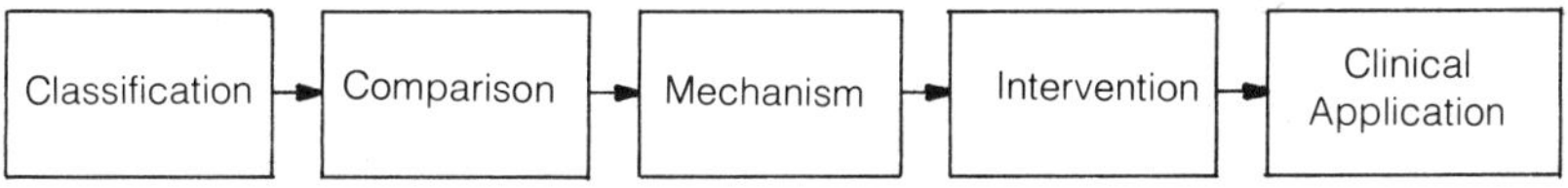

Classification → Comparison → Mechanism → Intervention → Clinical Application

FIGURE 7
EMPIRICAL RESEARCH LOGIC

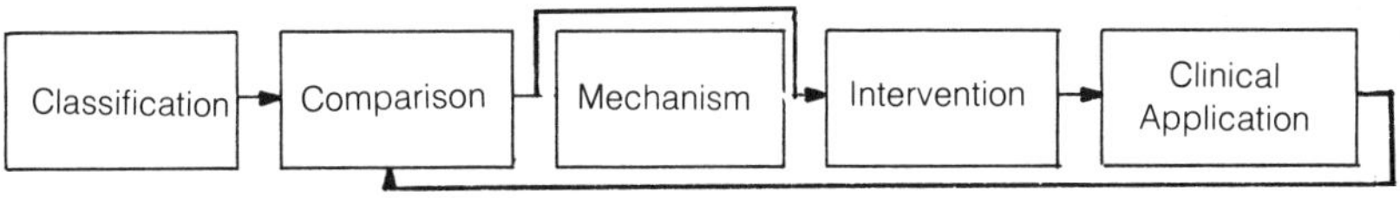

Classification → Comparison → Mechanism → Intervention → Clinical Application

Our model outlines a stage theory of the innovation process. Stage models of such processes are useful descriptions of analytically discrete steps, but lack power to assist understanding the actual process of moving through a stage[5] or from one stage to another. The process of sizing up the state of knowledge and selecting a suitable course of action never will be easy. Although this process is important at the start of a program, the research approach requires continual re-evaluation as a program generates knowledge. The ability to identify appropriate stages or to unfreeze existing inappropriate approaches and switch to new ones is important in managing technically successful programs.

Knowledge, the output of research, implies that learning is an important part of the process; learning models, such as Kolb's four-step repetitive cycle, have been used to describe it.[6] Learning as it takes place in the innovation process is not linear, but iterative. Often it becomes necessary to loop back through a previous stage to pick up knowledge applicable to newly arisen uncertainties and continue through the stages to completion.

A learning loop has four basic activities:[7]

Divergence: This activity is necessary in situations of uncertainty when problems and activities need identification. This process creates strategic variety which is essential if an organization is to cope with the uncertainty of its environment.

Assimilation: This activity builds on the structure created by the divergence process. Alternatives are compared, theories developed,

---
[5] "R & D organizations as Learning Systems"; B. Carlson, P. Keane and J. B. Martin. *Sloan Management Review*, Spring 1976.
[6] Ibid.
[7] Ibid.

hypotheses put forward and problem foci refined, and approaches outlined.

Convergence: Efforts are focused through choice among alternatives, hypotheses are tested, and decisions are made; all reduce further the uncertainty of the problem.

Execution: Objectives and schedules are set and implemented, and resources committed.

While the entire innovation process could be viewed as one learning cycle, it is important to recognize that each stage in the process also is characterized by uncertainties peculiar to it and requiring resolution prior to the next stage. The entire innovation pipeline then can be visualized as a series of learning loops within the larger cycle (Figure 8).

Proceeding through these loops is not automatic, but problematic. Managers must ensure that programs loop back to necessary learning activities when problems arise. A critical question becomes how does a research manager know when another step in the learning cycle is required. Lack of progress, emerging differences of opinion, changes in constraints such as resources, or other feedback indicating difficulties may be the only symptoms. The wise response is to stop and resolve the issue rather than ignore it.

The Millimeter Waveguide Program progressed sequentially from discovery through exploratory development, although not without difficulties and setbacks. It is a good example of a technically successful program that illustrates the stages of the R & D process, their complexity and the learning cycle concept.

The Millimeter Waveguide Program

Millimeter waveguide was a high capacity, high frequency transmission system for long-distance telephone communication. It carried voice, data, and facsimile signals in waveguide tube buried underground in a protective steel sheath.

It was the brainchild of Bell Lab researchers G. C. Southworth and S. A. Schelkunoff. Schelkunoff's theoretical work in mathematics progressed more or less concurrently with Southworth's physical experimentation. In 1933 Southworth telegraphed the first waveguide message through water-filled copper drainpipe mounted on fence posts. His message: "Send Money."

Waveguide was unique since signals were broadcast not through space, but through a single physical conductor. Practical application required extremely high frequencies. As frequency increased, the diameter of the waveguide decreased, but signal attenuation increased. Using mathematical theory, Schelkunoff discovered that, for a particular electrical field in circular waveguide, signal attenuation continued to decrease as frequency increased. A waveguide system was theoretically possible in a size potentially economical for use in long-distance transmission systems.

FIGURE 8
STAGES IN THE R & D PROCESS

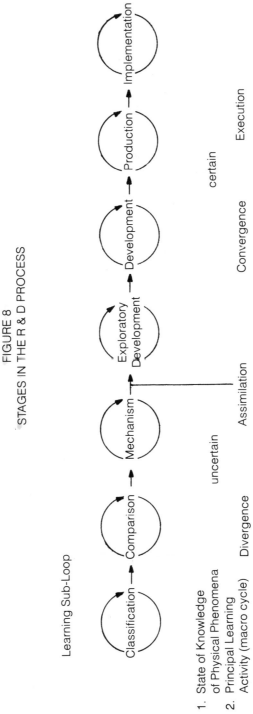

Southworth, writing in 1936, questioned the practical use of waveguide, however. The frequencies necessary for use were the highest tried for radio at that time, and the state of the art did not permit satisfactory evaluation of waveguide's applications.

The early waveguide work provided the basis for rapid development of radar, but it wasn't until after World War II that waveguide received serious attention as a communication medium. At this time, Bell Lab researchers began investigating techniques for encoding digital signals, having recognized that millimeter waveguide's frequency range was suited for digital transmission.

Most existing transmission systems and the majority of the Bell network were analog. The transmitted electrical voltage wave is the exact analog of the pressure wave created in air by the human voice. This complicated signal must be preserved to its destination. As it progresses through the system, its clarity fades and it must be amplified. Each amplification degrades the signal's quality.

Digital transmission obviated the need of preserving an analog of the original voice waves by converting the signal to a series of binary digits which could be regenerated during transmission and be reconstructed at the destination. Digital transmission, while not a perfect solution for all systems, was well suited to waveguide. Designers could work with extremely high frequencies, while retaining transmission quality. The practical application of waveguide was becoming more probable.

The First Exploratory Development Program

In 1958, Bell Labs authorized an exploratory development program to learn more about waveguide's capabilities and limitations. At this phase in a program's history, the design "ideal" was initially translated into a prototype and analyzed. The goal was to decide what could be built and at what cost.

Waveguide was in exploratory development for four years when it came into direct competition with coaxial cable to fulfill a system requirement. In June 1962, a team representing Bell Labs, Western Electric, and Long Lines decided that waveguide was the better medium, but that it had too many unresolved problems: Western Electric couldn't manufacture it without difficulty; solid state electronics were available for coaxial cable systems, while waveguide still used electron tubes; and demand projections didn't justify the capacity of the waveguide system. Coaxial cable was chosen and, in effect, the first exploratory development effort ended.

The report recommended that research on solid state electronics for waveguide continue. A small group of researchers continued this work and, in 1966, demonstrated the first solid-state, waveguide repeater.

The Second Exploratory Development Program

A month after this demonstration, the Transmission Council, composed of Bell System executives, established a committee to review all existing and future guided wave media. With radio space becoming a scarce resource, this committee was to examine alternative technologies for efficient use of this resource. After reviewing coaxial cable, waveguide and optical (laser) systems, the committee concluded it was time to resume work on waveguide. A proposal was made to build a twenty-mile section for analysis.

Designers produced a test model with parts manufacturered by Western Electric. Western Electric participated in design reviews, supplying its manufacturing know-how, and began to get experience producing component parts.

Manufacturing Waveguide for Testing

Creating a totally new transmission medium was a critical difference between this exploratory development project and most others. Bell Labs had extensive experience in designing new electronic systems for existing transmission media (paired wire, coaxial cable, radio) and testing them using spare cables or channels designated for future growth. Researchers could test and evaluate new equipment in existing operational systems. In the case of waveguide, however, the medium did not exist and had to be manufactured. It was an absolutely new product that no factory ever had produced. Even the manufacturing machinery had to be designed and built.

The exploratory development group worked closely with Western Electric's Engineering Research Center. ERC's responsibility in the Bell System was research into manufacturing processes. It was important that Western Electric be involved early in the program to develop the new manufacturing techniques which clearly were needed. Western Electric had to learn all it could about waveguide: how to make it extremely straight and how to apply a special coating inside the guide. A cooperative effort started when Western Electric sent two employees from its factory to the Bell Labs' development organization—with their salaries paid by Bell Labs. ERC also rented a small plant halfway between its facilities and the research lab as a pilot plant for the research. These two groups worked closely on quality and reliability issues, and measurement and testing techniques. After two years' work, characterized by long, hard interchanges about what could and could not be done, ERC developed a process for manufacturing waveguide.

It eventually was decided to contract with a steel tube manufacturer to produce the waveguide. This decision added a responsibility for tight coordination with the contractor, since the normal ways of specifying steel tube dimensions were inadequate for waveguide. Besides the diameter and straightness of waveguide, requirements included detailed geometrical specifications in terminology unfamiliar to the steel tube industry. Although the steel tube manufacturer couldn't make this measurement, Bell Labs and the ERC solved the problem by developing a computerized gauge to be pushed through the tube at different stages of manufacture.

This solution uncovered the next obstacle. Existing manufacturing processes met the required geometric specifications except at the final stage, which introduced an imperfection. ERC designed a new component for the contractor's equipment. Manufactured steel tubing then was shipped to Western's plant where various processes for plating and lining the inside of the tubing were tried and evaluated.

Eventually, specifications were written from which 11 miles of waveguide were produced for field evaluation. The project director commented:

> After five years of interaction, we'll agree on and write the final "specs." The reason is that we now know how waveguide works, and Western Electric knows it can design and build a factory to produce it. But even these aren't the final specifications for commercial production.

Performance was one objective, but reliability was equally important. Millimeter waveguide represented a very high front-end investment before the first circuit was operational and Bell could not afford to have anything go wrong after it was buried.

Installation Techniques

Long Lines Department of AT&T had to learn how to install waveguide. Installers were familiar with the relatively simple task of laying cable—open a trench, lay in the cable, and splice it. Unlike cable, waveguide's ultimate performance was directly related not only to mechanical imperfections, but to terrain and installation conditions as well. Bell Labs next worked with Long Lines to develop satisfactory installation methods.

Initially, designers considered developing laser-guided trenchers to ensure a level trench bottom. However, the Long Lines man on loan to Bell Labs had installed more than 10,000 miles of cable, and convinced them that special considerations for the trench bottom would be very expensive; they should stay close to standard pipeline-installation techniques.

In 1973, standard pipeline techniques became feasible and the final decision was made for that mode of installation. The invention of a special support by one of Bell's engineers made it possible. This individual understood the relationship between filters and electrical loss in electrical systems and was able to transfer the concept to a mechanical system. The support fit around the waveguide and not only buffered it, but also allowed the waveguide to be inserted into a sheath already buried. The support, acting as a mechanical filter, effectively filtered out curvature caused by the uneven

The Technical Logic of Research and Development

trench bottom and considerably reduced that source of electrical loss. The waveguide system now could be installed at low cost.

The invention came late in the program when Western Electric could not take responsibility for developing and producing it. The development group took responsibility and the inventor was assigned to work with an outside contractor to build a model. After four months, the prototype was ready and tested out well in the system.

Field Evaluation Test

The exploratory development proposal funded in 1969 included a test installation, and the first test measurements were taken in March 1975. The goal was to produce an operational test system which could be certified as functional and as meeting specifications.

Most of the components tested as expected, but when the entire system was tested, there was a surprise. At high frequencies, more signal loss was recorded than expected; an additional electrical field caused by the terrain that the theorists had missed was identified. There were two alternatives: get rid of the distortion or learn to live with it. The latter was chosen and Long Lines learned to deal this complication into route engineering.

Waveguide now was ready to leave exploratory development and enter commercial development.

Waveguide Put "On Hold" Once Again

By 1976, millimeter waveguide had completed exploratory development and field testing. It now was possible to manufacture and install waveguide systems, but again there was competition for use in Bell's long-haul transmission network. The combined effect of evolutionary advances in existing systems and in computerized network management caused the Millimeter Waveguide Program to be delayed and re-scheduled.

In the short run, a mix of these options permitted economical expansion. In the long run, some technological revolutions loomed large. Researchers were investigating satellite systems, and exploratory development work was being done on optical fiber systems which transmitted information by lasers.

The Millimeter Waveguide Program director summed up waveguide's history and possibly its future when he commented that millimeter waveguide might be always a bridesmaid and never a bride.

The Technical Logic of Millimeter Waveguide

The millimeter waveguide story began in the 1930s with two researchers working with different research methods. One discovered waveguide as a transmission medium through physical experimentation, while the other discovered the characteristics of circular waveguide using mathematical theory.

Both investigators were members of Bell Labs' Area 10 which conducted "fundamental" research or discovery for the Bell System. During the discovery phase, waveguide was a novel, laboratory phenomenon of possible use, but certainly beyond the technology of the time for practical application. Discovery took time. Ignoring the hiatus imposed by World War II, the total time involved in discovery was 10−15 years.

The technical logic of biomedical research can be compared with that of millimeter waveguide. There are similarities in knowledge generation in the biomedical realm of living systems and the electro-mechanical realm of inanimate physical systems. Terms such as "causal mechanism" and "clinical trials" are inappropriate outside biomedical science. But by moving to a slightly more abstract level, our seven-stage technical logic reflects reasonably well the waveguide experience.

In waveguide, the stage one process, classifying and categorizing properties, had been completed in developing higher frequency radio. Prior work at Bell Labs, and elsewhere, had mapped out the relationship between wave length and frequency. Starting with this knowledge, the two researchers, each in his own way, picked up the key idea of a physical waveguide and started doing stage two comparative studies of the interplay between frequencies, wave lengths, and physical guides. They mapped out those quantitative relations.

The special combination of these variables that would enable a message to be sent through the system was discovered through physical experimentation. This was the qualitative stage three that cannot be called a search for the "causal mechanism," but is roughly analogous in its search for a link between a complex phenomenon (guided radio waves) and a useful event (message transmission). In more general terms, this might be called a search for the *critical linkage*.

Establishing the linkage between radio waves and message transmission was the turning point that eventually permitted entrance to the fourth stage—exploratory development. The objectives of this stage were:

1. understanding the technical characteristics, capabilities, and limitations of waveguide.

2. translating the theoretical ideal into objectives for specific development, while ensuring technical quality.

3. learning to manufacture and install waveguide and understanding the faults and degradations that can occur during these activities.

4. understanding and being able to control the functioning of waveguide in its ultimate environment (context) and its interface with the existing physical network.

5. building a workable *system* to the point where the technology could be "sold" to developers who would refine it for production.

The detailed nature of waveguide and its limitations were mapped out, and the necessary equipment and sub-systems developed. As the whole system was shaped up, work also began on identifying and solving boundary or interface issues between the system and its varied operating environments. Finally, the project moved to the second, exploratory development, period of actual field testing which clearly parallels clinical testing.

Even at this stage of the process, a great deal of uncertainty was evident and the exploratory development group's primary control was knowledge—understanding the waveguide medium. They learned about the system in exacting detail by building models, and by actual implementation in a field test. Certification that the integrated test system was functional and met objectives was the conclusion of the exploratory development phase. The focus of this phase was on the whole system interacting with its environment.

Had waveguide been economically competitive with alternative systems, or if growth demands had been stronger, Western Electric would have assumed primary responsibility for the fifth stage of the innovation process, development, with continued engineering control and technical support provided by Bell Labs. This stage would have focused on achieving manufacturing economies through further refinement in product design and production processes. Then there would be two final steps for waveguide—larger-scale, low-cost replication (production) and dissemination of the system (implementation).

Millimeter waveguide provides an example of both an entire research program as a learning cycle, and also each stage as a separate learning cycle. Initially divergent approaches (theory and experimentation) at the beginning of waveguide in the 1930s complemented each other and provided a basis for formulating approaches to this new medium and for defining potential problems with certain electrical fields. To the Bell System, waveguide itself was an example of divergence, since the other long-distance, high-volume communications media such as coaxial cable already existed.

The convergence activity of the first exploratory development program ended, not in execution of a waveguide system, but rather in a decision not to continue. This decision was the result of further divergence: administrative, rather than technical. Alternative systems were compared, problems with waveguide recognized, and another system chosen for use in the communications network.

This instance shows that there is not only a technical learning cycle, but also an administrative learning cycle which integrates the former into the goals of the organization.

An administrative decision basically put the large loop into neutral while a complete sub-loop began. Solid-state electronics were needed as an alternative to vacuum tubes, and eventually were developed. This factor permitted the entire project to get back on track. Convergence continued with the second exploratory development program, which also had completed sub-loops for the manufacture of the waveguide medium and the supports necessary for its installation. The program successfully completed the convergence task but, again, the administrative learning cycle took control and postponed execution.

The waveguide program was technically successful: it met technical expectations and was economically feasible to produce and install. Bell Labs personnel clearly had articulated and understood the relationship of various stages in their technical logic, and progressed sequentially through these stages to completion of exploratory development.

Situations such as genetics research and millimeter waveguide make one think that the technical logic found here is inevitable. This is not so, as further cases will illustrate. What we found was that only by utilizing the type approach illustrated above can programs hope to achieve long-term success. Other programs were studied which appeared stuck in a single stage and unable to escape. These programs sometimes could achieve a limited success, but informed critical judgment in all cases indicated that greater success would have resulted from closer pursuit of the multi-stage logic and from a choice of stage appropriate to the state of knowledge about the physical phenomena.

Next, we examine two situations that suffered from faulty "stage management," both in the selection of starting points and in an inability to switch to more appropriate stages. "Politics" played an important role in shaping both the Artificial Heart and Cancer Chemotherapy programs.

Chapter 4

Linking the Political and Technical Environments: Two Unsuccessful Programs

Hopes and expectations often can reach unrealistic proportions. The day-to-day reality of research can be incomprehensible and boring to non-researchers. Supporters of research like to hear and talk about tangible end products like miracle drugs or a glamorous piece of technology. The hope provided by these products encourages actors in the political environment to project victory in the war against cancer or heart disease.

In both the Artificial Heart and Cancer Chemotherapy programs, discussed in this chapter, expectations exceeded reality. These programs were established and funded through action by coalitions of well-placed and well-meaning people. In both cases, the choice of a technical strategy was out of synchronization with the state of knowledge about the phenomena under study. In both cases, the result was an expensive program that did not produce the expected products. Both cases indicate that simply throwing a lot of money at a problem is not a guarantee of success.

It is not always the political environment external to the research organization that is responsible for side-tracking the technical logic. Research administrators and scientists can make this mistake as well. In the case of the Artificial Heart Program, the misassessment of the stage of knowledge and the choice of the wrong stage in which to start came from within NIH and not from Congress. As a consequence, once the idea of a transplantable heart was presented to Congress, quick moral and financial support ratified the program's judgment. Eventually, it was conflict over the appropriate stage which forced changes, not particularly successful, and resulted in a rejection of the program within NIH.

The Chemotherapy Program is an example of the external environment's dominating a research program and forcing a technical strategy on it. It illustrates what can happen when a program is a captive of a larger social process and is positively reinforced by liberal funding. The empirical strategy of chemotherapy, technically a calculated risk, was capable of absorbing available funds and providing a sense of movement against the dread disease. Eventually, it was a decline in financial resources that forced reconsideration of empiricism as the appropriate strategy. Its replacement, rational design, had as its basis the more traditional sequential strategy of knowledge development.

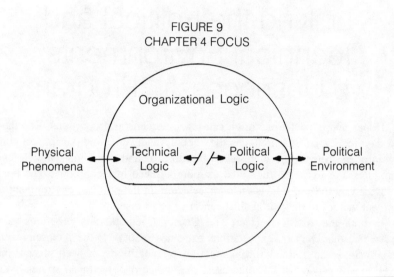

FIGURE 9
CHAPTER 4 FOCUS

The Artificial Heart Program

The possibility of replacing biological organs with mechanical devices dates at least as far back as 1939 to the development and use of a heart-lung machine; it continued through the next two decades with increased research on pacemakers, heart-lung bypasses, and various valves and pumps. Some scientists focused on developing auxiliary-assist devices, while others envisioned a totally implantable artificial replacement heart. By 1964, scientists working toward this latter goal estimated its fruition in 3 to 10 years.[1]

In November 1963, the Advisory Council to the National Heart and Lung Institute (NHLI) identified artificial heart research as a priority area. After this decision, the director of NIH urged the Heart Institute to hold a meeting to outline a broad program for artificial heart research. An *ad hoc* advisory committee composed of seven M.D.s and some of the institute's planning staff was convened. The doctors all were knowledgeable in cardiovascular mechanical-device development, having worked either internally, at NIH, or

---

[1] Testimony presented to House Appropriations Sub-committee for FY1965 by NHLI.

externally, on grants, toward developing an artificial heart. The group was convinced that partial assistance devices and even a total implantable heart replacement were desirable and feasible objectives.

The committee recommended establishing an office to coordinate artificial heart research. This office was the beginning of the Artificial Heart Program. It was to be a contract program using the capabilities of industry, but its objectives were vague. Whether the devices were to be the total replacement type or temporary was not made clear.

Also undecided about structuring such a program, the small core of staff held discussions with existing grantees to identify required hardware for which contracts could be issued. Nine contracts were issued prior to the end of fiscal year 1964 at a total cost of $581,000. Three were to study the interaction between blood and foreign materials and to develop blood compatible materials, while others were to fabricate pumps, drive units, and model circulating systems.

In 1964, developing an artificial heart was perceived largely as an engineering problem. Establishing the contract program with industry reflected a recognition that NIH generally lacked the critical engineering and production skills necessary for such an endeavor. Grantees or consultants to the Heart Institute were to draw up specifications for biomaterials or devices, for which the further development or production was to be contracted out.

Faith in the power of technology was in full bloom at this time. Satellites were being put into orbit, manned spaceflight was a reality, and the United States had embarked on a program to put a man on the moon. A technological euphoria existed. Not just the Heart Institute and its grantees held this view. An eminent Nobel Laureate in the field of genetics also saw the artificial heart as an engineering problem. In a letter to the director of the Heart Institute, he expressed his puzzlement as to why an artificial heart had not been made available as "a fruit of our technological developments."[2] He believed such a development was manageable within existing scientific knowledge and technical proficiency. The problem was that fragmentary studies were being pursued rather than a unified engineering program. "Its order of complexity," he stated, "should be similar to designing a sophisticated spacecraft and should cost around $100 million." A systems approach like that used by NASA was called for:

> . . . we do ourselves a great disservice to neglect the opportunity of a systems response to what is now a well-defined technical problem, which is so much a matter of engineering design, material development and empirical testing, and should not be confused with the basic research that was needed at its foundations.

---

[2] Dr. Joshua Lederberg. Letter to Dr. Ralph Knutti, Director National Heart and Lung Institute, dated June 24, 1964.

At this same time, the President's Commission on Heart Disease, Cancer, and Stroke endorsed the concept of an artificial heart. Although there had been no human heart transplants, open heart surgery was becoming routine and the scientific community recognized the possibility or even inevitablity of transplantation. It was unlikely that there would be enough donors for transplantation, and replacement hearts would ease that constraint.

By 1965, 17 types of artificial hearts had undergone preliminary trials in animals and none was considered sufficiently developed for human use. But optimism prevailed and the Artificial Heart Program escalated its effort. In mid-1965, the Master Plan was unveiled.

The Master Plan

Although there were dissenting opinions on the appropriateness of a concentrated systems approach and the perceived "engineering problem," this technological orientation was not displaced. Program staff were excited by the idea of a systems approach and pursued it through discussions with NASA and the Department of Defense which had extensive experience with this type of activity.

The artificial heart was conceptualized as a number of subsystems: blood pump, blood compatible materials, energy source, energy conversion system, and control system. Previously, researchers had been interested only in the blood pump itself and, if allowed to continue on their course, they would go through an evolutionary development process that would, opponents argued, "delay the achievement of an effective, workable, total unit."[3] A concurrent systems approach, it was held, would utilize time and money more efficiently than would the traditional sequential approach to research.

Program administrators had decided on what they denoted a modified systems approach comprised of a six-part cycle:[4]

i) defining objectives,
ii) designing alternative systems to meet the objectives,
iii) evaluating the technical and cost effectiveness of alternatives,
iv) questioning the objectives and assumptions underlying the analysis,
v) considering new alternatives and establishing new objectives,
vi) reiterating the previous steps as necessary.

The objective was clear: a totally implantable artificial heart. The master plan was the operational embodiment of what was seen as the remaining steps of the systems approach. It had four phases:

Phase 1: A conceptual phase in which the state-of-the-art would be assessed, design constraints identified through contracted studies, systems

---

[3] 1965 internal memorandum by Dr. Frank Hastings entitled "Background Material on the Artificial Heart Program."

[4] "Artificial Heart Program in Applied Health R/D" by Frank Hastings M.D. and Lowell Harrison Ph.D. February 1970.

specifications prepared, and program office staff hired. Phase 1 was to be completed by the summer of 1966.

Phase 2: The major activity was to be a "competitive effort of about four contractors resulting in specifications for the entire system including hardware, installation facilities, testing and training facilities, and a fixed price incentive proposal for the development of the artificial heart." This was to be completed by the fall of 1967.

Phase 3: By the fall of 1969, a number of prototypes would be designed and developed and the winner of the contract competition announced.

Phase 4: By 1970, the testing and certification of requirements for artificial hearts would be completed, resulting in the "availability of specifications to which artificial hearts can be mass-produced, installed, maintained, and monitored."

A target date of February 14, 1970, Valentine's Day, was set for achieving the program's goal.

Response to the Master Plan

By June 1965, a new *ad hoc* advisory committee endorsed the plan, and shortly thereafter it was discussed before the House Appropriations Committee. Congressmen were impressed. However, the director of NIH was not! He had agreed to neither the plan nor to the amount of money that the program was requesting—about $15 million per year. Beyond the administrative conflict, he was skeptical of the plan's feasibility.

First, he believed that the anticipated completion date was unrealistic—far too optimistic. Second, and more importantly, he thought the plan infeasible because of insufficient basic knowledge about functions such as the physiological rejection process. Third, he did not agree that an artificial heart was a priority item deserving approximately 10 percent of the Heart and Lung Institute's annual budget.

The conflict put the official implementation of the master plan in temporary limbo, but program activities continued in the direction that they had started. Concurrent with the completion of the master plan, six feasibility studies had been funded. Those were completed by the end of 1965 and a research organization was hired to analyze and summarize the six reports. This review was presented to the *ad hoc* committee consisting of M.D.s and Ph.D.s from inside and outside NIH, as well as some scientists from industry and the military. While there was disagreement on a number of key points and some doubt expressed as to the availability of basic biological knowledge to warrant a large-scale development program, the committee eventually recommended that the program be expanded. On the basis of the summary report and the committee's recommendations, the program issued six requests for proposals and effectively started Phase 2 of the master plan.

The director of NIH responded by cutting back the program's budget request from $15 million to $1.4 million; the internal conflict increased. Congress eventually raised the figure to $3.9 million, but the cut had constrained severely the program from pushing ahead with its aims.

In July 1965, the director of the Heart Institute retired and was followed by two short-term appointees until March 1966 when a new director took over.

After reviewing the plans of the Artificial Heart Program, the new director recommended a program with a dual thrust. One prong was the engineering program, while the other was basic research into myocardial infarction, or the reasons for heart failure. The latter thrust represented explicit recognition that insufficient knowledge of heart failure existed. The Director of NIH modified this proposal to reflect a greater emphasis on research into the conditions surrounding myocardial infarction and focused the engineering effort on developing partial, temporary-assist devices. The director of NIH pointed out that more knowledge was necessary about the cause of insufficient blood circulation, circulatory requirements under assisted conditions, the type of materials necessary to overcome problems of biological-machine interface, fluid dynamics of blood flow, and the development of energy sources.

The Myocardial Infarction Program was established with its own program manager to complement the engineering development effort of the Artificial Heart Program. However, in fact, there never was any integration of the two programs and they proceeded quite independently of each other, even though they were located on the same corridor.

The Artificial Heart Program tried to remain on its original schedule even when the lack of scientific knowledge and reduced funding seemed to argue against it. For example, the master plan envisioned the development of competing models of artificial hearts to be tested and evaluated. The need for a separate test center was based on the assumption that testing by the developer or its chosen medical collaborators might not be adequate or objective.

Each of thirteen proposals for evaluation centers was funded for six-month planning phases, after which one was to be selected as the test site. However, politics intervened and it became necessary to fund two geographically balanced sites. One was funded during fiscal year 1969 in the Mid-West and one the following year in the West. A substantial investment was required to create the facilities and hire the staff needed to test the artificial hearts. By 1973, the two centers represented about 20 percent of the program's budget, but had made little or no contribution since the expected completion date of a fully developed prototype had slipped to the late 1970s. Both centers eventually were closed during fiscal year 1974.

Although the original goal of the program was not met, some tangible results were produced. In 1969, the first total implantation of a heart-assist device was achieved in a calf. The device was, however, a long way from being used in humans. Development of power supplies continued, and in

1972, the first nuclear-powered left-ventricular-assist device was implanted in a calf. In 1976, human trials with blood pumps were conducted in postoperation, open-heart surgery patients. The goal was a temporary assist to give the natural heart time to repair itself. However, there is little doubt that the overall achievements of the program were disappointing and well below original expectations.

In September 1975, staff members of the Artificial Heart Program prepared a report for a task force convened to review program progress. They reported that even the basic design of the pump—size, location, shape, weight—still remained a problem. By 1976, $86 million had been spent and an artificial heart was still a long way off. Some people attributed the program's not meeting its goal to inadequate funding. Another view was expressed by the director of the Heart Institute in 1976:

> The science wasn't there when it started and it's not there today. The biomedical researchers expected too much from the engineers and vice versa.

The Political Environment of the Artificial Heart Program

The Artificial Heart Program arose out of a background of diffuse activities and scientific advances: developments in open-heart surgery and the prospect of heart transplants, artificial device development, analogy to technological accomplishments in other areas of science, plus some individuals' feelings that it was an idea whose time had come. These developments and beliefs were crystallized into a program by a few key actors.

The predispositions of individuals, their values and preferences, can exert enormous influence over the shape into which a program is molded. In this case, the key people were M.D.s with an applied orientation, rather than researchers primarily seeking knowledge. It also may have been significant that the original *ad hoc* advisory committee possessed a similar applied orientation. The committee members all were knowledgeable about mechanical-device development and had worked either within NIH or on grants toward the development of an artificial heart. This combination of people with similar viewpoints, interests, and possibly personal dreams, and occupying positions of influence made the original assessment about the state of knowledge in the field and the stage at which the Artificial Heart Program would begin its activities. The purported output, a piece of sophisticated technology, captured the imagination and political support of Conress and, temporarily, the prospect of the requisite financial support.

We have characterized NIH as a polycentric organization: that is, it is comprised of a number of interrelated power centers—the various institutes. The institutes derive their power from being funded as separate line items in the NIH budget and from being quite visible to Congress, the scientific community, and the public. Each develops its own set of supporters and benefactors. Since NIH is not a monolithic, hierarchial structure, the Office of the Director of NIH must be viewed as part of a program's political environment with its own constituency and power base.

This relationship was pertinent because the program and the director's office each represented different value orientations. The applied, developmental focus of the program ran counter to the mainstream pure-science orientation of NIH represented by the NIH director's stance. Program administrators appeared to want too prominent a place and initially revealed their desires in the wrong place. This act clearly violated the existing power balance within NIH and drew a sharp reaction—a dramatic cut in the proposed budget.

The value conflict might have been avoided by earlier intra-NIH discussion. The director of NIH, a supporter of pure research, had been involved in applied research during World War II and realized the value of such—in its place. But the objective and methods of the Heart Program assumed an almost ideological status and eventually the director was put in an adversary position. The director rejected the Heart Program as it stood and contributed to re-designing the program. The program manager rejected the new research thrust and continued in his chosen direction. This rejection process led to a condition of mutual isolation and the inability of either subprogram to learn from, and contribute to, the other.

By early 1967, the program was many things to many people: to the manager of the Artificial Heart segment, the program objective remained a totally implantable heart. The program had been underway for two years with a budget, contracts, and a staff. To the director of NIH, it was a more limited endeavor in which a new emphasis on basic research would provide the necessary understanding for the engineering development portion. Primarily, it was caught between two conflicting and unreconciled views of its nature. This mutual rejection and unresolved conflict became the major political relationship dominating and debilitating the program.

The Technical Logic of the Artificial Heart Program

The Artificial Heart Program, although not a major component in the Heart Institute's portfolio of projects, was clearly an initiative into the active management of a large-scale engineering development project. The technical logic was chosen largely by analogy to NASA space projects and a personal belief that the "systems-method" was appropriate for a large area of biomedical research. The central argument for the systems approach was that the then current investigators were interested only in individual sub-systems which had to be integrated to make the larger system a reality. Also, there were perceived dangers in developing half-way technology: temporary-assist devices which would facilitate survival through the myocardial crisis, but would leave the patient permanently bed-ridden.

While there may have been similarities with the NASA situation and experience, there were some fundamental differences. The system required to put a man on the moon can be visualized as a black box in its environment, the solar system. The structure and dynamics of the solar system form essentially a determinate system with known parameters. Although some uncertainty existed because of imperfect knowledge of the moon's surface and some concern about its effects on humans, the boundary between the black box and its environment was controllable. The environment was understood and uncertainties could be handled by providing defensive measures (space suits, decontamination precedures, etc.). The ability to isolate the black box for design purposes made possible a view of the development process as optimization within known stable constraints.

The artificial heart could not, and still cannot, be isolated as a black box problem. The program appears to have suffered from faulty stage assessment in choosing its starting point. The probable assumptions underlying this choice were consistent with a more certain, convergence-type activity. Such assumptions include:

(1) although not perfect, the state of knowledge is sufficient to support a project at the development stage;

(2) this knowledge now permits the designer enough control over critical constraints to create a functioning artificial device;

(3) the process of design or development consists of a series of iterations moving closer and closer to an optimization of the system within these constraints;

The leaders of the Artificial Heart Program began acting toward the technical task as if they were beyond the stage of discovering the causal mechanism or critical linkage between a certain phenomenon and a desired result. They felt prepared to specify the required sub-systems, and to begin integrated system testing under experimental conditions to be followed shortly by production of the devices.

Our judgment that the program began at the wrong point in the R & D process is based on an evaluation of the boundary conditions between that black box, the artificial heart, and its environment.

The discovery stage was not sufficiently advanced to support a *Stage 5* engineering development effort which artificial heart was assumed to entail. Patients for whom an artificial heart might be a feasible strategy were suffering from myocardial infarction, a condition which was poorly understood. The boundary conditions, or the interaction between the artificial heart and its environment, could not be explored adequately because of limited knowledge concerning problems in the arterial system, the reaction between biomaterials and blood, and the body's rejection processes. With hindsight, we can see that they really only were ready for *Stage 2* of our several-step process—ready for comparative studies and the search for the critical linkage. The Artificial Heart Program's defined task simply did not fit the state of knowledge of the physical phenomena.

Systems-thinking with related planning and control methods can be appropriate to large engineering development projects, provided the stage of knowledge of the relevant phenomenon is relatively complete about all subsystems, and about the relation between viable sub-systems and between the total system and its operating environment. When this stringent condition does not hold for the entire system or any part, the technical logic must loop back readily to an earlier phase and method. Systems-thinking is a useful tool in the appropriate circumstances—never an end in itself.

Although the systems approach which the program manager espoused clearly made provision for a feedback cycle or learning loop, the Master Plan, which was the logic in action, made no provision for actively incorporating the experiences of later phases so as to affect the assumptions of the earlier phases. Such a process was mentioned often, but never became a reality. In fact, "learning" seemed to be a process of eliminating competing approaches by finding them inadequate.

So far, we have used the term "systems-thinking" as it was used in the program, with its corollaries of planning, control and detailed master plans. A learning approach to systems-thinking would be different. Recognition of partial knowledge and partial ignorance could lead to an iterative cycle of divergence and convergence, and exploratory initiatives leading to new data and a new synthesis. This process might be characterized as systems-thinking and incremental action which would appear more suitable for situations with a clear design purpose, but partial ignorance of the natural phenomenon.

The Artificial Heart Program existed for more than ten years, despite several reasons for expecting its demise. To date, it has not achieved its original purpose; no artificial heart exists. It is possible to make a case that the program never was given a chance. The fact that the eventual funding level was not as originally planned provides a ready claim that the plan was undermined. It appears unlikely that the heart could have been produced with full funding, however. Cash resources do not appear to have been as significant a constraint as the lack of knowledge.

Possibly, the program never had a chance in another sense. It never appeared capable of learning from divergent viewpoints. Individual personalities may have blocked learning, and differences of opinion about the value of "basic" versus "applied" research possibly increased defensiveness. NIH's polycentrism apparently facilitated diversity, but could not ensure successful learning from that diversity. Conflicting views were neither reconciled nor synthesized to allow the program to move ahead on the basis of greater knowledge. In such situations of unresolved conflict, which can develop easily in the political environment of R & D programs, it is perhaps necessary to have someone who is committed to creating a synthesis. Unfortunately, in this case, that individual did not emerge.

Changing the organizational structure had little impact in this situation:

Linking the Political and Technical Environments 63

the organizational relations changed several times, but new learning never developed. Argyris (1972) is known for his skepticism regarding the impact of structural change on organizational learning. He emphasizes the need to confront the cognitive and emotional barriers preventing learning. The evidence of this case appears to confirm his position.

Comparing the Waveguide and Artificial Heart Cases

The state of the Artificial Heart Program was similar to that of the Bell Labs' researcher in the years before he had sent the first message through the waveguide. The missing linkage was between an artificial heart device and a patient in cardiovascular crisis. Until researchers had found, at least once, the combination of hardware and surgical methods, no matter how crude and temporary, that could serve as the circulating pump for a patient in crisis, they had not found the mechanism that could serve as the turning point for Stage 4 and 5 development.

It is, of course, easy to see how well-informed and well-intended people could have been mistaken about the development stage and choice of task logics. If an artificial heart device can pump on a test stand, why not when attached to a living being? The waveguide managers fortunately recognized that they had to invent and design the medium, as well as the black box, before they could move to exploratory development.

In contrast to waveguide, the Artificial Heart Program had difficulty maintaining the conditions that foster adaptation: the capacity to adjust the technical logic and program structure to changing the environmental conditions. Differing assessments of task realities, of systems planning, and of organizational capacity appeared early. These varied views might have led to new learning and synthesis if the political environment had been managed properly and opposing positions had not degenerated into righteous causes that were ends in themselves.

A comparison of the programs in terms of moving through the stages of our model is presented in Table 2. It summarizes the first five stages of our model and associated questions R & D managers should ask themselves. Had these questions been asked and answered, the Artificial Heart Program could have concentrated on much needed knowledge development.

In contrast to the artificial heart case, the people who shaped the Cancer Chemotherapy Program were aware of their limited knowledge about the disease. The program, an example of the empirical research strategy, was a response to pressure from the political environment to step up the fight against cancer. It was a calculated risk to find a cure quickly. Millions of dollars were spent annually to test thousands of chemical compounds, but without much success in developing miracle drugs. It highlights the fact that

TABLE 2
MILLIMETER WAVEGUIDE AND ARTIFICIAL HEART COMPARISON
TASK STAGE

PROGRAM	(1,2,3) DISCOVERY	(4) EXPLORATORY DEVELOPMENT	(5) DEVELOPMENT	RESULT
Millimeter Waveguide				Achieved Technical Expectations but not Implemented.
Question:	Do we understand the relationship between signal transmission phenomena and the new medium?	Q: Do we understand the ultimate environment (underground and the telephone network) into which this device must intervene?	Q: Do we have at least one way of producing a functioning system?	
Answer:	Yes —GO→	A: Yes	A: Yes	
		Q: Can we adequately control the interactions between the device/system and its environment?	Q: Should we produce the system?	
		A: Yes —GO→	A: No	

Artificial Heart

Question:	Do we understand the conditions contributing to heart attacks that precede the need for an artificial heart?	Don't Go!	Q:	Do we understand the ultimate environment (cardiovascular system, biological rejection processes) into which this device must intervene?
Answer:	No		A:	No
Question:	Do we understand what causes heart failure in those situations in which an artificial heart would be a viable solution?		Q:	Can we adequately control the interactions between the device and its environment?
Answer:	No		A:	No

Don't Go!	Q:	Do we have at least one way of producing a functioning system? → Never Achieved Technical Expectations.
	A:	No
	Q:	Should we produce the system?
	A:	Yes

throwing almost unlimited funds into an area of research is no guarantee of success.

Organizing to Fight Cancer

The National Cancer Institute (NCI) was established in 1937 as the first discrete institute within the National Institutes of Health. By 1945, there was a group of private citizens who were dedicated to fighting cancer. Key among this group were Albert and Mary Lasker who devoted considerable time and money in support of cancer research.

In 1944, the American Cancer Society raised $780,000, but spent none of it on research. In 1945, with the aid of Albert Lasker, a wealthy businessman, the society raised $4 million and agreed to spend 25 percent on research. In 1946, the Society raised $10 million, and Albert Lasker was named to the board of directors. The Laskers became very influential in this most visible and influential organization. After Albert's death in 1952, Mary Lasker continued to devote herself to fighting cancer and other diseases:

> For Mary Lasker, the conquest of disease was not just a slogan to arouse popular support. It was a real cause, a crucial and obtainable goal. After all, Mrs. Lasker says, polio, tuberculosis, and most other diseases of bacteriological origin have largely been conquered. "I'm really opposed to heart attacks and cancer and strokes the way I'm opposed to sin," she has said.[5]

After the war, the future role of government support for biomedical research was being debated among some scientists seeking to return to the poor, but pleasant, autonomy of the pre-war period and those seeking the excitement of growth possible with continued government support. The director of NIH at the time undoubtedly represented the former viewpoint in denying the need for funding beyond that requested when he appeared before the House appropriations sub-committee in 1946.

The Laskers and their supporters began developing plans for an alternative organization to NIH to carry out a national research effort. Before that plan materialized, however, NIH revised its position. Dr. Leonard Scheele, associate director for NCI, agreed that higher appropriations would be beneficial. In 1947, Dr. Scheele became the director of NCI, and NIH requested $26 million, with $14 million for NCI (compared to $1.8 million in 1946). In 1948, Dr. Scheele became Surgeon General; NIH had emerged as the focal point for biomedical research; and NCI had begun the growth which would make it the largest institute within NIH, with an annual budget ultimately reaching almost $1 billion.

---

[5] S. Strickland, *Politics, Science and Dread Disease*, Harvard University Press, Cambridge, Mass., 1972; p. 187.

A tacit, but highly successful, coalition eventually emerged among important Congressional leaders, Mary Lasker, and the new director of NIH. Strickland (1972) points out that although Senator Hill and Congressman Fogarty "had been influenced in the conduct of their responsibilities by Mrs. Lasker and her allies in the private sector; and their important friend, Surgeon General Leonard Scheele, had picked Dr. Shannon for the NIH directorship," there never was a formal alliance. However, we see once again the importance of a group of actors in a situation whose interests coincide and who became critical elements in the political environment of R & D programs.

Emergence of the Chemotherapy Program

Chemotherapy, the use of chemicals to combat malignant cells, is one of five basic methods for treating cancer. Although a relatively new method compared with radiation or surgery, it has been used with some success against systemic cancers, as opposed to solid, localized types.

Chemotherapy emerged around 1941 when reserachers reported that estrogen, the female sex hormone, was useful in treating prostatic cancer in men. Thereafter, a series of accidental occurrences reinforced the potential of chemical therapy. During World War II, some people were exposed accidentally to sulfur mustard gas destined for chemical warfare experiments and began to suffer a deficiency of certain elements in their blood. This condition eventually was exploited in cancer patients to deprive cancerous cells of the opportunity to proliferate. A related compound, nitrogen mustard, came into use in treating chronic leukemia and Hodgkin's disease. Later, it was discovered that people whose diets were deficient in folic acid produced a blood and bone marrow condition useful in treating leukemia in children. Finally, mass drug-screening in war-time antibiotic programs uncovered toxic chemicals displaying anti-tumor activity.

The Sloan-Kettering Institute, one of only three organizations in the United States having a combination of laboratory facilities for drug development and clinical facilities for testing, reoriented virtually its entire staff from the war-related program to a major peacetime chemotherapy program. The institute's director encouraged domestic and foreign chemical and pharmaceutical companies to submit materials to the screening program. By 1955, the institute handled 75 percent of the total chemotherapy screening activity in the United States. Sloan-Kettering reached its capacity, but the number of compounds submitted continually increased. Constant pressure from industry, academics, and laboratories for more screening capacity eventually resulted in the National Cancer Institute's assuming primary responsibility for drug-screening.

In 1953, the Senate, impressed by the results of the war-time antibiotics and antimalaria programs, encouraged NCI to establish a program for leukemia and added $1 million to its budget. There was little enthusiasm within NCI for such a limited scope program, so the staff took the initiative in structuring a program to expand the major existing chemotherapy programs into a cooperative effort. The Cancer Chemotherapy National Service Center (CCNSC) was established as the coordinating organization. The Cancer Society joined NCI as co-sponsors of the program and one of its representatives was named to the advisory committee. Although the program was not exactly as prescribed, the Senate was satisfied with the direction of events and happy that some visible action was to be taken.

The strategy for the program had to be decided and two classical viewpoints emerged: the "empirical" versus the "rational" approach. The rational or sequential approach was to design or select drugs for testing on the basis of knowledge about the way in which a drug worked and about the natural history of cancer cells which permitted developing hypotheses about the effect of a drug on a tumor.

The empirical approach meant random screening of large numbers of drugs in search of "leads" for further exploration. This strategy treats both cancer and the drugs as black boxes, devoting no attention to the way either of them works. Rather, researchers observe relations between the drugs and quantitative changes in tumors.

Empiricism, never seriously challenged, was chosen as the program's strategy. The program got underway and never looked back—especially since funding proved to be no problem. From an initial budget of $5 million, appropriations rose to $30 million in 1960, $42 million by 1965, and eventually exceeded $75 million in 1975. The number of chemical compounds submitted for testing reached 50,000 and more than 350,000 tests were being conducted annually. Approximately 20 of the tens of thousands of drugs submitted each year reached the point of serious consideration for clinical trials and an average of only about 7 was permitted by the Federal Drug Administration for clinical trials in humans.

The program grew in organizational complexity along with its increased funding, and from that point of view it was successful. But it had its critics. A continuing, but little heeded, criticism was the choice of the empirical strategy and, not unrelated to this choice, the very observable ratio of resource input to drug output. This measurable relationship provided perhaps the most damning criticism of the program. Chemotherapy was absorbing resources out of proportion to resulting benefits. Too much money was being spent for too little output!

The situation eventually changed in 1972 when the chemotherapy program was incorporated into the Division of Cancer Treatment (DCT). The focus of treatment was shifting from specific methods like chemotherapy, radiation, and surgery to the diseases themselves and how a combination of treatments

Linking the Political and Technical Environments 69

(multi-modality treatment) affects them. DCT spearheaded this effort.

Perhaps more importantly, financial resources were cut back and new ways had to be found for testing drugs. The rational approach gained strength and, by 1976, a selection process for drug testing was put in place to reduce the input to about 15,000 "theoretically" chosen drugs. The head of DCT commented:

> The recent reorganization of the DCT symbolizes a move towards "rationalism." The reasons for this changed direction center around two things—increasing costs and increasing numbers of compounds to be tested . . . These constraints necessitate the generation of criteria for choices and basically this demands the rational approach whereby one pursues families of drugs which appear to have some generic relationship with each other and one or more of whose members has been seen to have some relevant biological activity. Another way of progressing is to work at the mechanism level and to investigate why it is that certain types of molecules have certain types of effects on cellular structures.
>
> It becomes possible under a rational strategy to investigate the way in which this natural process works and to seek concrete ways of facilitating it.

The Political and Technical Logics of Chemotherapy

The Chemotherapy Program provides another example of the political environment dominating the technical strategy of a research program. The initial choice of an empirical approach fitted the circumstances of the time. Interest groups such as the American Cancer Society were pushing for more research on cancer. Congress, encouraged by the wartime success in antibiotics, convinced NCI to increase its activity and provided liberal funding. The more money provided by Congress, the more drugs were tested—visible evidence that the money was needed. The vicious circle of growth, depicted in Figure 10, blocked for some time the learning which would have shifted the technical logic earlier.

The values around the traditional scientific paradigm of a sequential approach were being weakened by the heavy funding available after the war. More important, as with the artificial heart, was the choice and use of an analogy in the argument for the proposed program. The space program legitimated an engineering development program using the systems approach in artificial heart; the antibiotics program legitimated the empirical approach for chemotherapy. These analogies not only legitimated the chosen approaches, but also patterned communication. The workings of any complex field tend to be explicable only in highly technical language which is difficult for lay people to comprehend. The success of the space and antibiotics programs was a metaphor meaningful to all interested parties. The trick, obviously, is to know when the metaphor is appropriate. That means knowing something

FIGURE 10
THE VICIOUS CIRCLE OF THE CHEMOTHERAPY PROGRAM

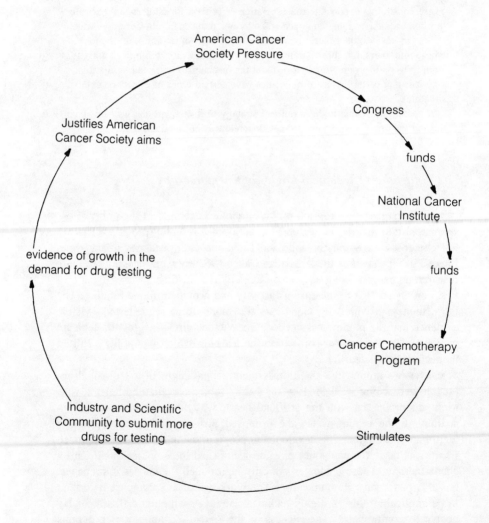

Linking the Political and Technical Environments 71

about the technical issues. One research scientist at NIH provided, simultaneously, the empiricist's viewpoint and a reason why the analogy to antibiotics may not have been appropriate:

> The general criticism against the empirical chemotherapy program is that, given the large amount of funding, not too many agents have emerged of great value in the clinic. I am not sure that this is a warranted criticism—if the field is not ready for major advances you will not get them. Empirical research will be the mainstay for the next two decades because we don't know enough about the biology of cancer. We don't know enough about the cause of cancer to design rationally. There is nothing wrong with the empirical approach. All the antibiotics were picked up that way. You cannot put all your eggs in one basket. Of course there is a biological difference between bacteria, against which antibiotics are effective, and cancer cells. Bacteria are one-celled, fairly simple organisms which have cell walls. The walls are broken down by the antibiotics and the body's natural immunological process wipes out the cell. In cancer the cells are mutant "normal" cells and have no cell walls in the same way. The cells in cancer are much more complex.

Unfortunately, from a cost-effectiveness perspective, all the eggs had been put in the chemotherapy basket for a long time. There undoubtedly is a place for the empirical strategy, given the right type problem and provided it doesn't get out of hand to consume unwarranted resources and block out other approaches.

Empiricism was a political and administrative success. The assumption of empiricism, as defined and used in NCI, was basically, more is better. The way to generate therapeutically useful agents is through testing more of them. This requires more facilities and resources. Empiricism, therefore, was the strategy which allowed the program to survive and prosper at a time when a problem was having too much money to spend. If the program had had a rational paradigm, it would have had no means of spending such money and it would have had serious political problems. Empiricism provided an infinitely expandable way to spend money and a quantifiable measure of activity and commitment, however spurious. It failed because it did not produce. Re-evaluation of the technical logic resulted in a multi-modality focus and rational testing displacing empiricism.

In the Artificial Heart Program, the task logic employed a four-step sequence as shown in Figure 11.

FIGURE 11
THE TASK LOGIC OF THE ARTIFICIAL HEART PROGRAM

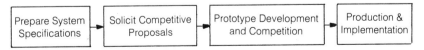

Basically, this is the way you would act with Stage 3 knowledge and the belief that the artificial heart was primarily a design problem. The Chemotherapy Program did not attempt to design a solution, but rather ran tests against some initially established screening criteria methodically to isolate potentially superior anti-cancer agents, as shown in Figure 12. We would suggest that these two models can best be seen as alternative ways to carry out the exploratory development phase of our model. If the mechanism that is the critical causal link between biological phenomena and a disease entity is well understood, then the systems model might well be a relevant approach. If the mechanism is not understood, the empiric model can be considered as a calculated high-risk strategy in attempting to achieve a solution without the benefit of a complete discovery phase. Utilization of either approach, however, should be based on knowledge of inherent risks and of the potential drain on resources.

FIGURE 12
THE TASK LOGIC OF THE CHEMOTHERAPY PROGRAM

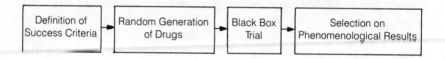

We can see, with the benefit of hindsight, that the managers of the Artificial Heart Program assumed the state of relevant knowledge at the start of the program to be essentially through the mechanism stage and, therefore, available to support a major engineering effort. Subsequent history has not borne out this assumption. In the Chemotherapy Program, there was a greater awareness of the lack of mechanism knowledge even as the empiric research was undertaken. This was at least partially acknowledged as a higher risk strategy; therefore, it should have been periodically reviewed.

While the task of each program had unique features that would make it unwise to try using the seven-step model in a simple literal way, it can be seen as a general framework to guide the choice of technical logic. Our study clearly highlights the importance of choosing the appropriate strategy, not only at the start of a program, but also through re-evaluation as the program generates knowledge. It seems that this realization finally dawned on the people running the "War on Cancer" and on the Congress which ardently supported it as the excerpts from recent *New York Times* stories illustrate.

Linking the Political and Technical Environments 73

The process of sizing up the state of knowledge and selecting a suitable strategy is never an easy one even without the interference that can be created in the political environment. The challenge is to manage the interplay between the political and technical environments. It takes a lot of energy, drive, and hard work to manage this interface and to make a research program function correctly, as we discovered in the Sudden Infant Death Program described in the next chapter.

RESEARCHERS PLAN MAJOR POLICY SHIFT
IN FIGHTING CANCER
by Harold M. Schmeck Jr.
(Excerpt from the N.Y. Times; May 22, 1978)

In a major policy shift, the National Cancer Institute is backing away from the view that the time is ripe for a tightly planned and blueprinted research war against cancer.

Instead, according to Dr. Arthur C. Upton, director of the institute, the agency is returning to the view that cancer, as a grouping of diseases, represents one of the central and most complex puzzles of biology and that the research strategy should be based on wide-ranging investigations.

The idea of a highly organized, meticulously planned cancer research effort has been prevalent among government officials during much of this decade. Although many denied that the objective was comparable to that of the Apollo program, which put men on the moon, the comparison was often made.

A Bicentennial Present

Enthusiasm for that approach led to proposals for a rapidly expanding budget that would reach $1 billion a year before the end of the decade. A few proponents even suggested that it might be possible, given enough money and effort, to conquer some major forms of cancer by 1976 as a Bicentennial present to the nation.

"There are clearly areas where programming is necessary," said Dr. Upton in a recent interview. Among these, he included areas in which the institute sees the need for major, expensive applied research programs.

"But, on the other hand," he added, "there are still vast areas of resarch where we are unable to lay out a blueprint and timetable."

5 SENATORS QUESTION INVESTMENT
OF U.S. FUNDS IN CANCER STUDIES
by Richard D. Lyons
(Excerpt from the N.Y. Times; June 14, 1978)

Five Senators demanded to know today whether the billions of dollars invested in cancer research had been properly spent and, if so, why so little progress had been made in seven years of Government-sponsored study.

The Senators, led by George McGovern of South Dakota, a Democrat, and Robert J. Dole of Kansas, a Republican, closely questioned for more than two hours the two scientists directing the Government's campaign against cancer. While the queries were very polite, the Senators repeatedly sought the kind of elaborations and specifics that Congress has been reluctant to pry out of the scientific community until recently.

A Growing Restlessness

The questions raised today only underscore what an increasing number of members of Congress, of both houses and both parties, regard as the public's growing restlessness over the lack of tangible results in a national effort to conquer the disease.

"I have a suspicion that we're losing the war on cancer because of mistaken priorities and misallocation of funds," Senator McGovern said, adding pointedly, "There has been no lack of funds—it's almost $1 billion a year."

Today's questioning of Dr. Donald Fredrickson, director of the National Institutes of Health, and Dr. Arthur Upton, director of the National Cancer Institute, at a Senate subcommittee hearing also demonstrated that Congress is becoming less enthusiastic about providing the nation's biomedical research community with a virtual blank check on Federal funds, under the 1971 act that opened the cancer campaign, without a better accounting of where the money is going and why.

Chapter 5

Linking the Political and Technical Environments: A Successful Program

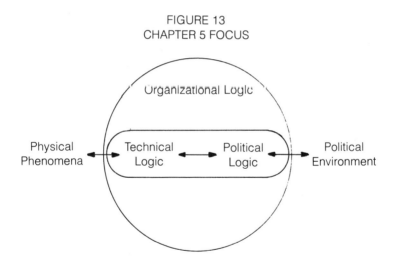

FIGURE 13
CHAPTER 5 FOCUS

The Sudden Infant Death Syndrome (SIDS) Program described in this chapter stands in contrast to the programs discussed in Chapter 4.

The state of knowledge about the syndrome was assessed accurately; a realistic and flexible program was set up; and some positive results were achieved. At times, the political environment threatened to overwhelm the program's technical logic, but through the personal commitment, hard work, and skill of the program manager, it never happened. Her concern for good science and her sensitivity to the needs of groups in the political environment allowed her to remain in control of the program and protect its integrity.

A symbiotic pattern of mutual adaptation eventually developed between the SIDS Program and the scientific community, special interest groups, Congress, and the administration of NIH. This condition, a mutually beneficial relationship, can develop when neither the program nor its constituencies is unilaterally dominant.

Sudden Infant Death Syndrome (SIDS)

The Sudden Infant Death Syndrome, or crib death, is the leading cause of death in the United States among infants one to twelve months old. Each year, between 7,500 and 10,000 infants die from SIDS. The syndrome generally occurs in infants who, although possibly having signs of a cold, are apparently healthy and can eat without difficulty. The infant, placed in its crib for a nap or for the night, is later found dead. Death is sudden and unexpected.

Normally, these deaths occurred at home and, if the children were taken to a hospital, it was usually only to the emergency room for attempts at resuscitation. The typical case, therefore, never penetrated the hospital's diagnosis and treatment system. The physiological conditions contributing to SIDS were unknown. This lack of knowledge was compounded by the fact that the starting point in any investigation was a dead baby, previously an apparently healthy child. Even after an autopsy, no cause of death could be found. Nor was there a mechanism for collecting data on SIDS death, since the syndrome was not classified in the World Health Organizations' categorization of causes of death.[1] Prior to 1976, at least for administrative purposes and statistical analyses, an infant could not die from SIDS; deaths had to be attributed to some other condition.

SIDS was essentially a private, family problem about which nothing was known. This secrecy tended to increase the emotional consequences for the parents, particularly the mother, who often blamed herself (and was blamed by others) for being negligent and causing the death. Public ignorance exacerbated the situation and, in some instances, even led to arrests for murder.

The social isolation and lack of knowledge helped create a vicious circle which reinforced the *status quo*. The scarcity of recorded incidents kept researchers from focusing on the syndrome. Since science generally was accepted as a self-stimulating activity, meaning the scientific community defined what it considered to be the important research topics, the lack of re-

---

[1]The SIDS program has changed this and the 9th edition of this listing published in 1976 contained SIDS as a cause of death.

search interest produced few funded grants and a minimal research effort and thereby reinforced the condition of no knowledge of SIDS. This vicious circle is illustrated in Figure 14. For parents and other laypeople, ignorance of SIDS led from false assumptions and attribution of guilt for the deaths to withdrawal and silence which also contributed to SID's remaining a mystery.

The National Institute for Child Health and Human Development (NICHD) was established in 1963. It supported some ongoing scientific activity in SIDS, but it was fragmented. No organized, formal research program began until 1971. From 1964 through 1971, NICHD funded a maximum of three primary grants or contracts for SIDS in any year. A "primary" grant or contract addressed SIDS directly, while a "subsidiary" one covered research activity in infant death which may have been related to SIDS, but SIDS was not the principal focus. NICHD also sponsored two conferences—one in 1963 and one in 1969. One well-known early researcher in SIDS commented that not until the 1969 conference was it "finally possible to say that sudden infant death syndrome was a real disease entity that is readily definable and not some vague mystery killer."[2]

NICHD failed to interrupt the vicious circle primarily because of its self-view—basically passive in relation to the definition of research areas. This role reflected the values of the *laissez-faire* approach to science, holding that the scientific community should be self-stimulating and autonomous, and that administrative agencies should not interfere or attempt to redirect the efforts of science. The culture of NIH also was one of response to articulated problems, rather than of proactive articulation of them. Such conditions prevented NICHD from breaking the circle of events.

Activity in the Political Environment

Other events during the decade of the 1960s, outside the scientific community and NIH, would be felt by NIH. Organizations of parents who had lost children to SIDS began to form.

The first parent organization, the Mark Addison Roe Foundation, was founded in 1962. Senator Lowell Weicker of Connecticut, a friend of founders Mr. and Mrs. Jedd Roe, was a member of the original board of trustees. State Senator and Mrs. Dore of Seattle, Washington, founded the Washington State Association for Sudden Infant Death Study, which later merged with the Roe Foundation to become the National Foundation for Sudden Infant Death Syndrome. Mr. and Mrs. Saul Goldberg founded the other major parent organization, the International Guild for Infant Survival. These groups were committed to support research aimed at understanding and pre-

---

[2] Dr. Abraham Bergman, Hearing before the Senate Subcommittee on Children and Youth of the Committee on Labor and Public Welfare, Jan. 25, 1972.

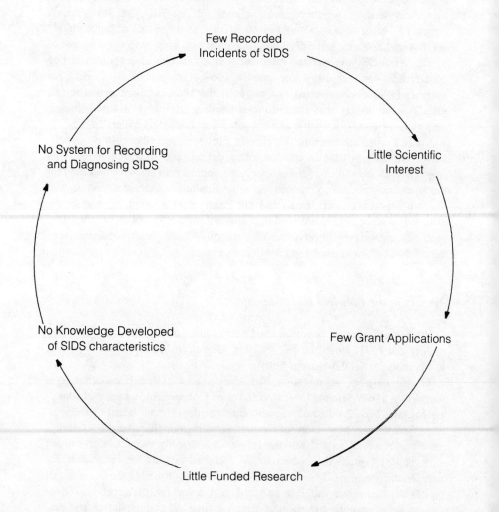

FIGURE 14
THE VICIOUS CIRCLE SURROUNDING SIDS

venting SIDS, and to educate public authorities so that grieved parents would be treated sympathetically and helpfully.

In addition to the emergence of such special interest groups, state and federal governments became involved in the SIDS problem. In 1963, the Washington State Legislature passed a law permitting autopsies on all cases of sudden, unexpected infant death to be performed at a university-affiliated teaching hospital and provided $20,000 to initiate a research project. In Oregon, Senator Packwood, whose friend's child died of SIDS, sponsored legislation funding SIDS research at the University of Oregon Medical School. In 1966, President Johnson directed that a specific infant mortality program be established. NICHD responded with a task force to examine activities and opportunities in the area and accepted its major recommendation for a Perinatal Biology and Infant Mortality Branch (PBIM) within the institute.

When Dr. Eileen Hasselmeyer assumed the leadership of PBIM in 1969, the situation was as described earlier—there was no organized research activity on SIDS. However, there was a growing recognition, inside and outside NICHD, that further initiatives in SIDS research were necessary.

The parent organizations were becoming disenchanted with NICHD and were at least partially responsible for the increased awareness of the unsatisfactory situation. NICHD may have contributed to the hostility and emotionalism developing among the parent groups. Representatives of these groups, invited to the 1969 conference at their own expense, left convinced that much more could be done. Parents' letters to the institute were answered very formally by a staff member, and the parents were kept at arm's length. Finally, the director of NICHD invited one parent to visit him, but it was too late. By that point the parents had developed the attitude that NICHD should go to them rather than them going to NICHD.

Initiating a SIDS Program and Sizing-Up the State of Knowledge

Dr. Hasselmeyer and her assistant, Jehu Hunter, became the prime movers in formulating a research policy for SIDS. NIH management, under increasing pressure, told them "do something, anything." In April 1971, a new, expanded SIDS research program was born.

As Dr. Hasselmeyer and Mr. Hunter began to outline the program, they asked themselves some questions:

> Jehu and I pondered questions such as: Where were we in our understanding of why SIDS happened? What new knowledge was needed to predict or prevent SIDS? What did we know about the underlying mechanisms of SIDS?
>
> It soon became evident that a new approach was needed and unless new blood was transfused into our approach, meaningful inroads into our understanding of crib death would not be accomplished.

The first activity of the new program was to be a research planning workshop. Dr. Hasselmeyer and her co-chairman for the workshop took a novel approach to inviting participants. They invited only a few individuals considered knowledgeable about crib death and concentrated on identifying top-notch scientists working in disciplines possibly pertinent to SIDS. Dr. Hasselmeyer knew she wanted representation from neuro-physiology with an emphasis on sleep physiology and maturation of the autonomic nervous system, neonatal cardio-respiratory physiology, immunology and infectious disease, epidemeology, and pathology, as well as one or two individuals knowledgeable about animal models for the syndrome.

One "great authority" in SIDS thought this type of workshop was the wrong approach to the problem; reliance on experts in SIDS was the "right way." The conventional wisdom in the planning stage of research programs, according to Jehu Hunter, said they should consult with the experts, ask them what needed doing and what would be feasible given the state of the art. However, Dr. Hasselmeyer was not satisfied with the results the experts had produced during the previous eight years and decided to continue with her plan to create divergence by "mixing insiders and outsiders."

Her perspective on the problem differed significantly from the experts: she saw the problem as being related to child development. Healthy babies just do not die without cause; there had to be antecedents in the prenatal stage, she reasoned. It was this failure to see SIDS as a developmental issue that was leading pathologists to imagine that, because they had examined the respiratory system, they knew all there was to know. Even when outside consultants recommended a separate SIDS branch be established, she prevailed in keeping it within PBIM as an integral part of the larger problem of infant development and infant mortality.

The first workshop, held in August 1971, was attended by 13 persons, many of whom never had participated in SIDS research, nor considered the relevance of their work to SIDS. The emphasis was on the developmental, physiological, and environmental phenomena which conceivably could contribute to the syndrome. As discussions progressed, the importance of adopting a multi-faceted, integrated, inter-disciplinary approach to the SIDS mystery became evident. Participants began pinpointing relationships between their specialties and SIDS. Dr. Hasselmeyer was elated: "The infusion of new life into our approach to SIDS had begun."

One of the chance discoveries of the day centered on sleep physiology. Everyone at the meeting, except the sleep expert, assumed children slept like adults. Children fall rapidly into a deep sleep while adults progressively enter deep sleep. This discovery laid the groundwork for the first grant in sleep physiology on live infants—research which proved to be partially applicable to SIDS. At the same time as this workshop, Dr. Hasselmeyer also was announcing that she had funds available for research. The announcement was printed on the back cover of the program for an annual meeting of a pediatric society.

After reading the transcript of the workshop and preparing a summary report, program personnel realized they had only scratched the surface. They identified six emphases and started planning workshops in each of them. Although most of the workshops focused on physiological aspects of SIDS, Dr. Hasselmeyer thought that the psychological problems experienced by parents, relatives, and community associated with a sudden infant death also deserved attention. This also became a workshop issue.

During three later workshops, numerous questions were asked about specific pathologies of SIDS cases. The inability of pathologists to answer them refuted a scientific consensus of the first workshop that the pathology of SIDS was complete. A pathology workshop was scheduled to switch activity in this discipline back to an earlier stage in the R & D process.

The workshop series succeeded in stimulating and challenging the participants. Dr. Hasselmeyer had begun to reverse the traditional, non-interventionist role of NICHD, at least in relation to SIDS. Several attendees later began work on crib death and received financial support for their research.

Managing the Political Environment

Congress Becomes Involved with SIDS

During 1971, parental concern about SIDS increased. The parent organizations began a successful telephone and mail campaign, deluging the entire executive branch and Congress with letters claiming inaction on SIDS. Needless to say, these letters ultimately found their way to the program administrators who spent 50–75 percent of their time for a year responding to them.

During this campaign, the parents were extremely vocal. Dr. Hasselmeyer recalled that NICHD was accused of lying and of clouding the issues by claiming the supplementary grants were related to SIDS research. Parents could not understand how a study of sleeping infants was relevant and they wanted the SIDS program to fund every grant application, good or bad, that had the words "sudden infant death syndrome" in them.

Congress, in January 1972, called NICHD to testify about its SIDS activities: primarily to answer why more was not being done. At that time there was one primary grant of $46,000 and numerous supplementary grants totalling $900,000. Apparently Congress was attentive only to the level of primary grant support and was upset. The director of NICHD and the assistant secretary (HEW) for health and science represented NICHD. Dr. Hasselmeyer and her colleagues from the program did not have an active part in the hearings. The traditional non-interventionist role of NIH was emphasized, while the SIDS planning workshop and other program activities were not. NIH, which appeared to be dragging its heels, made a poor showing compared to the parent organizations, which introduced harrowing testimony and factual evidence showing SIDS to be the most important single cause of infant mortality.

The message from Congress was loud and clear—do more research on SIDS! Congress began legislating to mandate certain activities. By August 1973, six bills and three resolutions had been filed in the House of Representatives. There was then a real danger that political forces and events might overwhelm the SIDS Program as had happened in the sickle cell anemia case.

Congressional hearings were held again in August 1973. This time, the program was presented well with Dr. Hasselmeyer's and Mr. Hunter's assistance. Congress was told about the series of imaginative, scientific workshops and the resulting identification of emphases. NICHD initiated activities, such as announcements in 30 scientific journals expressing an interest in funding SIDS research, were breaking the vicious circle that blocked scientific progress. The $46,000 expenditure on primary grants of 1972 had increased to more than $400,000. Not only was more money being spent, but identification of some areas where knowledge had accumulated and where new leads were being pursued also was possible. This additional funding for SIDS came, not from external sources, but from shifts of priorities in existing budgets. Following the 1972 Congressional hearings, the director of NICHD and Dr. Hasselmeyer reviewed PBIM's priorities and funding and made changes.

The final SIDS Act was less directive regarding the SIDS program than discretionary in tone. The one mandate was that an annual report be presented to Congress. The Act did not appropriate additional funds for the program, although there were suggestions in political circles of earmarking $10 million for it annually.[3] Dr. Hasselmeyer fought against additional funding for the SIDS program:

> I drew up contingency plans and suggested ways that the money could be used, but it was clear that we were against this approach.
>
> I believe in paying only for the best research. That way you increase the possibility of getting reliable findings. If you have an earmarked budget, then you may be funding some poor research which will confuse the subject and delay progress. You need to have flexibility. The fewer constraints in funding that we have, the more flexible we can be in pursuing opportunities and changing direction as necessary.

---

[3] The actual expenditures of the program during the period FY 1972 and FY 1976 fluctuated between $3–5 million dollars.

Linking the Political and Technical Environments

Dr. Hasselmeyer prevailed and continued to manage the program according to her philosophy and with her entrepreneurial spirit. Her 1976 annual report to the National Advisory Council of NICHD explained that there was a growing body of evidence indicating that crib death was not the inexplicable mystery it once was believed. Researchers had discovered that the victims were not completely normal, but rather had subtle physiological defects of a neurological, cardio-respiratory, and/or metabolic nature. It also had become evident that the cause of SIDS was multi-factorial and not the result of a single mechanism.

By 1976, the problem of SIDS had not been solved, but as one official at NICHD expressed it, "We are light years ahead of where we were."

Interacting with Parent Groups

Comprising a significant constituency were the parents and their organizations. During the 1971 mail campaign, when the negative attitude toward NICHD was developing, Dr. Hasselmeyer and Mr. Hunter decided that they would attend a parent organization meeting. Dr. Hasselmeyer got to know the parents and began a very positive relationship with them. She commented:

> After that meeting, I wrote a nice, chummy, lady-to-lady letter to the wife of the organization's founder whom I had met. She wrote me the nicest letter back, thanking me for attending the meeting.

Dr. Hasselmeyer and Mr. Hunter helped the parents understand SIDS and what was being done about the problem. It paid off, as Dr. Hasselmeyer indicated:

> We went out at night to meet them and talk with them. It was a real concentrated effort that neither of us particularly wanted to undertake, but we did because it was important.
>
> We began to get to know many of them and gradually they came around and saw the relevance of what we were doing. It was a real reversal—from thinking they were being deceived to believing us.

Every time a SIDS workshop pamphlet was printed, Dr. Hasselmeyer sent pre-publication copies to the organizations with a hand-written note. They often wrote back asking for 100 copies. Since that time, relations with these groups and individuals changed dramatically according to Dr. Hasselmeyer:

> Now they are very friendly. We receive Christmas cards and pictures of their children. Now they want to know what they can do for us like raising money or putting announcements in their newsletters.

In addition, Dr. Hasselmeyer counseled some SIDS parents. She told them to call her if they had a problem and needed to talk. She was pleased with these accomplishments, but noted that she and Jehu Hunter did it all on their own, working nights after full work days.

> We had no institutional support at all. The director didn't put up any roadblocks, but he didn't help us either.

The parents knew more about SIDS and were convinced that a serious research effort was under way. Their expectations were brought in line with reality. Dr. Hasselmeyer not only diffused a hostile and potentially overwhelming set of special interest groups, but also gained a supportive ally for her program.

Interacting with the Administration

HEW and NIH's Office of the Director also were key groups for the program. In these relationships, the use of an inter-agency committee and the Operational Planning System were key elements. The SIDS Program administrator informally organized an inter-agency coordinating committee to keep her informed about SIDS-related research in other parts of NIH and HEW. The assistant secretary of health and a number of parents were concerned with the potential lack of coordination between SIDS-related activities in NIH. The assistant secretary had pledged the formation of a coordination group, but upon hearing it already existed informally, he gave it official status. The committee increased support for the program and located the areas where interesting and related research was being carried out. Proactive behavior of this sort by Dr. Hasselmeyer helped the program to retain its initiative and integrity.

During the Nixon Administration, a push to introduce more professional, businesslike administration into government introduced a system of establishing program objectives and monitoring progress, termed the Operational Planning System (OPS). It met resistance at NIH because of the ingrained attitude against "planned" science. It was used only reluctantly and under duress. Each institute had an obligation to the secretary of HEW to place two programs on OPS. SIDS went on OPS in 1972, against the advice of NICHD.

Progress was reported on a monthly basis. No reported problems produced a "satisfactory" evaluation; a delay was rated as a "minor problem"; and two minor problems constituted a "major problem." A rating of "major problems" usually meant the institute director had to complete a variance analysis explaining how the problem was being resolved.

This system, which many scientists rejected because they sensed an attempt at programming, was used by the SIDS program in a unique fashion within NIH. It was used flexibly.

The first year, a number of priority areas was indentified in an OPS document. Changes after the first workshop necessitated redoing the OPS plan. The prevailing attitude in the NICHD and NIH hierarchy was that the original objectives should remain and work progress toward them. The SIDS program administrators refused to let the plan dominate their activities and convinced people that their approach to OPS made the system viable. The core of their argument was: "Why should we work toward objectives that have become obsolete? The SIDS program is advancing rapidly." They also refused to schedule and program scientific breakthroughs: only administrative milestones were scheduled.

The following year, SIDS had a "major problem" which the director of NICHD had to explain in person to the assistant secretary of health who, it turned out, was quite happy with this flexible approach to OPS.

In 1974, during a program review with the director of NIH, Dr. Hasselmeyer discussed her use of OPS. The director was impressed and began using the program as an example of what could be accomplished within an existing budget, and as an example of OPS use. A year later, Dr. Hasselmeyer met with the then-acting director of NIH on the subject of OPS. These meetings normally took ten minutes and often were viewed as dreaded experiences. Dr. Hasselmeyer saw it as an opportunity to provide a positive image of the SIDS program, and the meeting lasted forty-five minutes.

Commentary on the SIDS Program's Interactions

Program personnel devoted considerable time to interacting directly and indirectly with the political environment. They educated the parental groups about the program and the utility to the SIDS program of various kinds of activities, particularly the administrative categorization of supplementary project fundings, originally a cause of confusion. This interaction also identified areas in which the program could produce useful educational materials and, not insignificantly, it served to express sympathy and provide some comfort for seriously grieving parents. The program administrators demonstrated an active receptivity to the situation of the parents and the meaning SIDS had for them.

The receptivity was selective. There was respect for parental concerns, but not total acceptance of their proposed solutions. Funding all projects with SIDS in the title would have been detrimental to the integrity of the SIDS research process. Effective responsiveness, therefore, required selectivity, which itself required some articulated criteria about the nature of the research process and its continued viability. These criteria were a concept of the program's role in the larger infant mortality problem and an understanding of what constituted "good" science.

While the parent groups were managed in a highly personal way, PBIM related to the scientific community with a mixture of personal activity and more formalized procedures. The idea that SIDS was a disease like others and must have symptoms was converted into a number of action priorities.

These were used as guides in defining the research workshops and the type of research to stimulate and support. The mix of projects changed over time as new dimensions of the problem were discovered and so, in parallel, did the research workshops. These formal mechanisms for determining the state of knowledge and possibilities for research were supplemented by Dr. Hasselmeyer's personal style. She cultivated an extensive network of contacts and was meticulous in the attention she paid to maintaining it; she spent a great deal of time on the telephone.

OPS was a great success because it was used as a tool to link the program more closely to the NIH administration and executive branch at a time when it needed visibility and support. The intelligent use of this tool provided the SIDS program with important support and the administrative hierarchy with valuable information on progress in SIDS.

Each component of the political environment which could affect the effectiveness and efficiency of the program was handled with different mechanisms:

1. HEW and NIH—OPS, Program Review, Interagency Committee
2. Internal (NIH) Scientific—Interagency Committee
3. External Social—Letters, meetings, educational materials, telephone networks
4. Congress—Congressional hearings
5. External Scientific—Workshops, informal task network, normal grant procedures, etc.

The degree of formality, complexity, and importance of these mechanisms matched that attached to them by the relevant group or by the desire of program personnel to influence and change their environment. In each situation, program administrators learned the concerns of environmental constituencies and simultaneously taught them something about the program's progress. The response used a format meaningful for those groups and directly addressed their concerns with demonstrable action.

Contrasting the SIDS, Artificial Heart, and Sickle Cell Programs

SIDS Program administrators successfully differentiated their environmental focus. There can be little doubt that although this process is vital to the success of any program, it is not inevitable, as we saw in artificial heart and sickle cell anemia. The artificial heart and sickle cell anemia programs each failed to fulfill the learning and teaching activities necessary to link themselves to their political environment in a beneficial way.

Administrators of the Artificial Heart Program chose their development objective and systems approach largely by ignoring opposing views. Having once convinced themselves that they were right, they made another serious mistake by dealing directly with Congress on budgeting matters without prior discussion and agreement with the director of NIH. This was an easy trap, however, since the expectations of Congress and many scientists were so high. If a man could be put on the moon, surely an artificial heart could be built. The beliefs and expectations of the program administrators were congruent with those external advocates of an artificial heart and therefore the administrators could not reduce these external expectations. Failure to learn and understand the position of the director of NIH created the unresolved conflict from which the program never recovered. Resources were spent on development projects that should have waited until more was known about the entire circulatory system and about heart disease. To the director of NIH, the Artificial Heart Program was only one of many programs needing support and its unilateral action endangered the entire Heart Institute.

While the administrators of the Artificial Heart Program primarily failed to learn, the administrators of the Sickle Cell Program failed to teach its constituency groupings what was feasible and what was not. It seems evident that the NIH administration knew what a program like the proposed Sickle Cell Program had to look like—primarily education and testing once reliable tests were designed. But rather than explain to HEW what was possible and how NIH could contribute, they avoided the issue until program leadership was thrust upon them.

The program was overwhelmed by special interest groups and politicians who meant well but lacked understanding of the disease, trait, or the research necessary to provide relief for people suffering from sickle cell anemia. The expectations of these groups were not brought in line with reality; in fact, it took four more years to achieve this result. Perhaps many unfortunate incidents could have been avoided had people at NIH initiated the interaction with their political environment.

The Technical Logic of SIDS

The SIDS Program had been in existence for a relatively short time during the course of our research and little can be said about the technical logic in relation to the program's ultimate success or failure. However, looking at the initial activity of sizing-up the state of knowledge and choosing a program strategy is very instructive.

When Dr. Hasselmeyer took over PBIM, nothing was known about SIDS. It was a mystery. In the face of scientific uncertainty, she wisely concluded that a discovery level effort was essential, and began creating a situation of divergence by exploring a variety of disciplines and viewpoints for potential relevance to SIDS.

Under Dr. Hasselmeyer's leadership, the initial requisite for structuring the SIDS research activity was that the syndrome could not be viewed in isolation from the encompassing problem of infant mortality. The guidelines used to design the scientific workshops—"infuse new blood" and "don't rely on the experts"—helped break down old boundaries and permitted a fresh exchange of information and ideas. New boundaries were drawn tentatively around seven problem areas rather than around disciplines. In areas such as the pathology of SIDS, where Dr. Hasselmeyer found knowledge to be less than originally assumed, she acted to loop back through the learning cycle to pick up that knowledge.

The program objectives provided a broad, flexible outline for establishing more specific goals. The first three were oriented primarily to the discovery phase of the technical task:

- To increase the understanding of underlying mechanisms of the syndrome.
- To discover its cause.
- To stimulate the scientific efforts toward finding the solution to SIDS.

The next seven reflected a concern for increasing knowledge of the syndrome and for specific action within the constraints of existing knowledge:

- To identify infants at risk of becoming a SIDS victim.
- To explore preventative approaches.
- To inform the scientific and lay community about the sudden infant death syndrome through scientific publications and public information documents.
- To provide training of *biomedical* and *behavioral* scientists.
- To support interdisciplinary conferences and workshops concerned with the sudden infant death syndrome.
- To learn more about the current status of management of SIDS cases in the United States.
- To develop guidelines for use by coroners, medical examiners, and pathologists in handling SIDS cases.

The last objective clearly focused on issues important to a critical component of the program's political environment:

- To clarify the impact of a sudden and unexpected infant death on the subsequent behavior of parents, siblings, and the extended family.

Although the SIDS program was primarily a discovery level effort, activity appropriate to later stages of the R & D process was initiated within existing financial and knowledge constraints.

Program management carved out areas of interest to its wide range of constituents, set objectives in those areas, and proceeded to perform in them. This balanced approach was quite successful and in 1976 the objectives were being re-evaluated internally and revised, because knowledge of SIDS had increased and the field of research expanded. Dr. Hasselmeyer was able to achieve a measure of success, because she maintained the program's scien-

Linking the Political and Technical Environments 89

tific integrity by requiring good scientific research and by proactively managing the program's political environment.

Administrative Mechanisms Supporting the Technical Logic

Ultimately, any idea for a research program must be executed: the ideal translated into action. The translation takes place through use of administrative mechanisms like task forces, committees, and schedules. These mechanisms are used to plan and initiate a program, to control it and monitor progress, to change direction as necessary, and to disseminate results. There are a standard array of these mechanisms available for any manager's use, some of which we will discuss briefly. However, surveying these is less important than understanding the relationship between the way they are used and the learning needs of a program at any particular time. These managerial tools, depending on their use, can be either aids to adaptiveness or traps. The choice and use of these administrative mechanisms should be consistent with the stage of the research program.

Although many administrative devices could be cited, the important issue is not the specific mechanisms, for they may be limitless, but the role they are intended to serve. The general requirements seem to be either to increase variety (divergence) or to decrease variety (convergence). In uncertain circumstances, environmental variety needs to be mapped into the program, and the mechanisms serve this function. When the task is more certain, variety needs to be reduced so that it doesn't overwhelm the program, allowing it to proceed through the latter stages of the technical logic as quickly and economically as possible.

The decisions to increase research on SIDS or artificial hearts, for example, became operational through group processes. The initial planning workshop on SIDS was instrumental in the formulation and planning of the program. It was successful in providing new ideas for NICHD's approach to SIDS. At inception, the Artificial Heart Program used an *ad hoc* committee for a similar purpose, but achieved a markedly different outcome. Why did the planning group mechanism get one program started on the right track and the other launched at an inappropriate stage? The composition of the two groups was undoubtedly a substantial factor. In artificial heart, the "experts" on the committee who had been working on aspects of artificial heart development increased convergence around the development of hardware rather than identify the true learning needs of the program and outline a program to address them. Continued use of committees blocking diverse views and master plans resembling PERT charts locked the program into an inappropriate stage, increased its insularity and resistance to change. Each step removed the program further from the modes of behavior appropriate to challenge the assumptions on which it was founded.

The planning workshops for SIDS, on the other hand, identified areas where knowledge was lacking and broadened the program's perspective, rather than narrowed it. They created the variety of ideas and approaches program management needed to begin focusing research support and setting objectives, and they called into question some initial assumptions.

In each of these situations, the use of the planning committee mechanism was consistent with the program manager's perception of what was required. In the case of SIDS, it was also consistent with the state of knowledge about the medical problem.

NIH also has two principal ways of providing R & D funds—grants and contracts—which probably are useful under different conditions of certainty. Grants originate with individual researchers and are more open-ended, thus providing a way to increase variety through divergent approaches to a problem. Contracts assume convergence has occurred and the remaining requirement is development of a specific end item. At NIH, there are study sections comprised of scientists from outside NIH that review grant proposals and rank them according to scientific merit. For a first-time proposal, the ranking functions as a resource allocator; but for researchers seeking continuation funding, project achievement is reviewed, as well. Again, such a group can increase either convergence or divergence primarily depending on its composition. If reviewers favor fundamental research or discovery over more applied research, then proposals of that type may be favored over others as our discussion of the Genetics Program in Chapter 7 illustrates. Although an institute's advisory council makes the ultimate decision on the research portfolio mix, priorities established by study sections are quite influential.

Once programs are initiated, someone must monitor and control their activities. Again, the mechanisms selected to do this must fit the stage of learning and certainty of the situation. Under conditions of certainty, mechanisms like PERT, critical path analysis, and master plans are appropriate. In the Artificial Heart Program, the master plan was consistent with the technical logic which assumed greater certainty than actually existed, but was not synchronized with the learning requirements.

In chemotherapy, the controlling mechanism was the "linear array" production line for evaluating and clinically testing drugs. It was a highly routinized and regulated process, with various monitoring and decision points. As a production line for testing drugs, its efficiency probably was beyond question, but again the technical logic and its mechanisms were out of touch with task realities. The linear array became the embodiment of the empirical technical logic and its very efficiency inhibited new learning.

In the Bell System, it was not until most technical uncertainties had been resolved that scheduling of a deterministic nature appeared. When systems moved into production, detailed planning became possible and extremely important. At this stage, deterministic, closed-system mechanisms can be employed fruitfully.

This does not imply that planning and monitoring of program activities are impossible under conditions of uncertainty. The subtle difference is that under conditions of relative technical certainty, progress toward completion of the piece of technology that is the end product can be more easily quantified, programmed, and measured.

The way Dr. Hasselmeyer used OPS is an example of the more appropriate way to plan and control program activities under conditions of uncertainty. OPS was a system for specifying program objectives, scheduling milestones for each objective, and recording progress toward them. OPS met serious resistance at NIH because of the natural reluctance of people there to plan science. It was, therefore, used reluctantly and under pressure.

As initially conceived, all objectives and milestones had to have time and cost schedules. These proved difficult for grant programs like SIDS that were dependent on investigator-initiated proposals and that could not accurately forecast funding levels from year to year. It took 18 months to reconcile this issue, but it was resolved; no longer was it necessary to cost-out objectives and milestones. The milestones Dr. Hasselmeyer scheduled were administrative events which, when completed, should help the program move forward in resolving the SIDS problem. She did not attempt to schedule scientific or technological results. An example of her use of this system is shown in Table 3.

The monitoring, or review, mechanisms should help identify the need to change direction or revert to an earlier stage of technical logic, as we saw happen with the pathology component of the SIDS program, and even simultaneously to readjust funding patterns.

The Chemotherapy Program for a long time was locked into its mechanical drug testing process without learning from the growing evidence of its limited effectiveness, regardless of its outstanding internal efficiencies. Using the technical logic of empiricism to absorb the abundance of funds pressed on the program meant that the technical logic could not respond to emerging task realities.

Eventually the growth cycle ended when external organizations began imposing funding constraints. The resulting tension led the program finally to review its activities and develop new criteria and approaches to learning within new funding constraints. Too much money is not good. A positive, creative tension seemed to be induced by limiting funding levels.

Similar over-funding began to develop in the Sudden Infant Death Syndrome Program, but was avoided astutely by the program director in order to protect program integrity. Funding low-quality research would confuse the program's learning process with poor research findings. Thus, what on the surface might appear to boost a program's chance of success, a large influx

TABLE 3
OPS FORM FOF THE SIDS PROGRAM

National Institutes of Health	Operational Planning System
Organization: National Institute of Child Health and Human Development	Objective and Operating Plan Fiscal year 1973
Objective No. 1: To expand and intensify a research training program aimed at solving the problem of sudden infant death syndrome.	Resources Required $3 million PHS Act Section 301 Overall Evaluation ☐
	Status Report for Months of

MILESTONES	COMPLETION DATE
	J A S O N D J F M A M J
1. Complete series of SIDS Research Planning Workshops to develop specific research objectives.	1
2. Prepare summary reports from SIDS Research Planning Workshops.	2
3. Publish and distribute annotated bibliography on SIDS.	3
4. Publish summary reports from SIDS Research Planning Workshops.	4
5. Develop Requests for Proposals (RFPs) based on summary reports of SIDS Research Planning Workshops in areas of developmental immunology, cardiorespiratory and thermometabolic phenomena, developmental neurophysiology of sleep, and epidemiology as related to SIDS.	5

6. Develop RFPs on the psychological impact of SIDS on affected families.
7. Issue and advertise RFPs developed in Nos. 5 and 6 above.
8. Establish panels for review of contract proposals submitted in response to RFPs.
9. Review of submitted contract proposals by review panels.
10. Review of submitted contract proposals by Institute Contract Review Committee.
11. Negotiation of contracts as approved by Institute Contract Review Committee.
12. Evaluation of staff programming efforts to stimulate project grants related to SIDS.
13. Assess effectiveness of total expanded and intensified SIDS research program.

of funds, can be detrimental if it either forces the program to absorb too much variety before it is capable of doing so, or if it locks a program into an inappropriate technical strategy.

At Bell, we saw review mechanisms used as Dr. Hasselmeyer used them. The waveguide program is a typical example where funding oscillated through a review of technical progress and the needs of the Bell System. For example, major status reviews by vice-presidents of Bell Labs, Western Electric, and AT&T were held typically in support of a funding request. All organizations involved in the waveguide program reported on accomplishments and plans for future funds. When a new, long-distance transmission medium seemed needed, limited exploratory development funds were made available. When coaxial cable was chosen over waveguide, funding was cut to the level required to develop its sold state electronics. On completion of this task, and coinciding with the need for a high volume, long-distance transmission system, the waveguide program was funded for a new, exploratory development effort. The pattern of program funds' oscillation within the larger system's needs appeared to be consistent across the programs we looked at.

As the need for waveguide's capacity was pushed further into the future, funding levels decreased and funding intervals grew shorter (typically one year instead of two or three); as a result, the more frequent program reviews permitted a decision to stop.

Finally, we would mention an administrative mechanism that can play an important part in determining a program's success—leadership orientation. While this may seem an inappropriate item to include as an administrative mechanism, we do so to point out that, from an organization's viewpoint, there is a degree of choice and substitutability. Management should try to fit the orientations of its technical leaders to their tasks. What follows is a brief description of two distinctive leadership orientations—the technique champion and the program champion.

Frank Hastings–Technique Champion

Dr. Hastings was an M.D. in his mid-thirties when he joined the Artificial Heart Program in mid-1964. One of his first tasks was helping to develop the "master plan." He and another program administrator talked at length with people from NASA and the Defense Department about the philosophy of systems thinking, which came to rule the program.

Dr. Hastings became acting head of the Artificial Heart program in June 1965, the same time the director of NIH rejected the master plan and cut the budget request which constrained Dr. Hastings from pushing ahead with his aims.

In June 1966, the dual program idea emerged and Hastings was made head of the artificial heart component. To Dr. Hastings, in 1967, the pro-

Linking the Political and Technical Environments

gram was a systems-based effort to create a totally implantable heart; all other activities were meaningful only in relation to this objective. To the director of NIH, the program had not started; the actions taken during the fall of 1967 were to "initiate" a more normal NIH program of research with a developmental element concerned with incremental pieces of equipment useful in temporary, partial-assist situations.

The two areas were supposed to function as complementary thrusts towards a joint goal but, in fact, worked in almost total isolation. The head of the myocardial infarction branch remembered Dr. Hastings and the climate in the program:

> Hastings was a dynamic man. He was a surgeon in his thirties and was fond of the analogy of the space program. He was terribly ambitious and devoted to the heart program. He took initiative over the program's activities and staffing and growth with great independence of the institute. He was continually broadening the mandate of the program to suit his own ambitions; even after the director had reduced the objective to an assist device for partial replacement, Hastings regarded the funds going to myocardial infarction as a direct drain on the available fundings for the device activities. He was the only person who knew everything that was going on and he liked to keep everything to himself. My office was at the other end of the corridor from Hastings, but I saw little of him and knew little of what was going on.

Here was an intelligent, dedicated individual who perhaps became trapped by a dream to develop a piece of hardware long before the state of knowledge permitted. In retrospect, we see the misfit between an individual's desires and orientation and his situation.

Presumably the reverse of this vignette exists as well, although we did not come across it. However, it is not difficult to imagine a research-focused person heading up a development effort and impeding satisfactory progress because of a desire to incorporate continually the latest scientific advances. This individual might be characterized as a "research champion."

We did, however, come across another orientation, which concentrated on solving a program's problems within the context of the overall technical logic. This orientation seemed particularly functional in helping a program make the transition from a predominant research focus, through the trials and setbacks of exploratory development, and into development.

Bill Warters–Program Champion

Bill Warters received a PhD in physics and worked in research for five years before becoming involved in the first waveguide exploratory development program. He was highly enthusiastic about the role of discovery and its contributions:

> It's been worth its weight in gold. We know this basic research is valuable, although we're not at all certain about the time frame. We do know, however,

that at some time in the future, we will need to work at higher frequencies since that's where the available radio space is.

When waveguide lost the competition with coaxial cable, Warters was promoted to department head in charge of eight researchers who continued work on the solid-state electronics, while the other research engineers were transferred to optical transmissions (lasers). Almost ten years later, Warters remained in contact with this group:

> We keep in touch. I'll demonstrate waveguide to them and try to explain all the development problems I've run across to help them out. They keep trying to convince me that optical fibers will beat out waveguide. But waveguide is basically developed and I know how long it takes to develop a workable system.

After his group built the necessary solid-state components, Warters made presentations to his research colleagues, outside groups, and development people. Try as he might, he could not encourage the development of equipment using the components he had assembled. People kept telling him to go away.

Frustrated, he sat down with his boss and reached the conclusion that, although they had built all the elements, they had not built a complete system. They decided to build a system and Warters changed his coming year's planned work effort:

> I told my people that they were going to be developers and damn it, we were going to build a repeater. They didn't particularly like the idea, but they accepted it. We had originated the concept, but hadn't carried it far enough. Researchers, themselves, sometimes have to move slightly in the development direction in order to sell their ideas.

Bill Warters was pleased with the final program results and what he characterized as "that sometimes difficult interaction with Western Electric around the manufacture of the waveguide medium." He commented:

> We put a lot of emphasis on understanding waveguide in exploratory development, rather than focusing on developing electronics and tight specifications for the guide. There is a real tradeoff involved if you're contracting something like this. If the specifications are too loose, you won't get the quality you need. The steel people will meet them, but they won't beat them. If they're too tight, the guide becomes overpriced.

He summed up the program in these words:

> There's no question that advances in competing technologies have been responsible for delaying implementation and installation of millimeter waveguide. The date for the first installation has been re-scheduled from 1978 to 1984. We're nearly as far away from installing the system as we were when we started in 1969, and we certainly don't have monopoly on the future. Optical fiber, which may be the technology of the future, could become a serious competitor. Millimeter waveguide could end up always being a bridesmaid and never a bride.

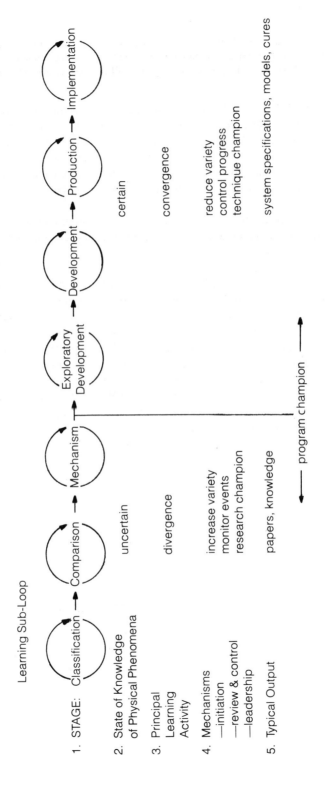

FIGURE 15
STAGES IN THE R&D PROCESS

Pushing the schedule out on waveguide was personally disappointing, but not disappointing from a Bell System viewpoint. If it's not economical, it's not right. If we can't develop something that will be cheaper than what exists, we don't do it.

In his behavior and attitudes, we see a concern for the problems of the total system—technical and economic performance—and an ability and willingness to loop back through some necessary learning stages. He was able to identify these needs and to make the necessary shifts. The program champion appears to be able to straddle multiple orientations and has the capability of switching the technical logic of a program as the situation dictates.

Figure 15 summarizes the stages of the innovation process and integrates the role of the administrative mechanisms that appear to be associated with successful performance.

Program leadership is important in managing an organization's research efforts. This is particularly true at NIH. NIH, a polycentric organization, experiences difficulty in transferring the learning from positive experiences, such as the SIDS Program, to other programs. Learning is left largely to the individual program managers. It is localized learning and dependent on research administrators' interests and skills. This type of situation, localized adaptation where the organization is almost synonymous with the individual, stands in contrast to the more institutionalized procedures of the Bell System. A major difference between NIH and Bell is that adaptive behavior responsive to changes in the technical environment of large R & D programs is institutionalized as an integral part of the system. We now will look at how Bell organizes and manages its technical activities.

Chapter 6

Organizing the Technical Logic

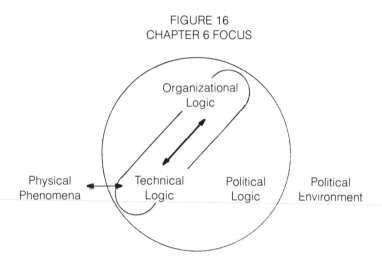

FIGURE 16
CHAPTER 6 FOCUS

As we observed in Chapter 1, the political environment often exerted influence at NIH for more applied research, and development programs. The potential danger, illustrated by the sickle cell anemia, artificial heart, and chemotherapy cases, is that the existing state of knowledge of the physical phenomena may not be sufficient to warrant such later-stage activities. Substantial resources may be channeled into programs, desirable as their objectives may be, that are unlikely to achieve the hoped-for results. Even without this knowledge limitation, however, the staff of research institutions like NIH may not have the predisposition, experience, or organizational support necessary to manage development programs. Serious organizational problems experienced in the sickle cell anemia and artificial heart programs hindered their progress. Knowledge transfer between people working at different stages of the technical logic was unsatisfactory and the activities of these people were not coordinated.

In two NIH programs, SIDS and genetics (discussed in the next chapter), managers articulated a concern for both generating and using knowledge, as well as a willingness to bridge the interface between researchers and users. Good science and a certain level of social responsiveness apparently were important values. But those managers also were pragmatic. They appreciated the related needs of people and organizations beyond their programs' boundaries and recognized the potentially disruptive impact these groups could have on their programs.

These two programs were relatively small, however, and did not require complex administrative mechanisms. An individual or a pair of individuals with this special orientation managed them effectively. But can this happen in large, complex research programs? Is it reasonable to assume that program managers as individuals can successfully reproduce and transmit this complicated set of behaviors in situations where hundreds of researchers, engineers, and administrators have important contributions to make and must be coordinated? In complex programs where a factoring of tasks creates a high degree of differentiation or specialization, organizations need to rely on more than the skill and charisma of individual leaders to provide the required transfer of knowledge and integration of activities.

The program manager's orientation undoubtedly is important, but it needs organizational support. This support ensures that all relevant information will be considered to minimize the possibility of mismatches developing between the technical logic and the state of knowledge. It also assists the transfer of knowledge through the R & D stages, and is critical in coordinating the activities of large, complex programs like the ones we observed at Bell Labs. One of the strengths of AT&T is the way it organizes and manages its technical logic. This chapter illustrates how the company uses its organization to produce technology and communication systems.

The Bell Laboratories Organization

Bell Labs is a mission-oriented research organization structured as shown in Figure 17.

Areas 10 and 20 are the science and technology generators of the Bell System. The research area (Area 10), oriented primarily toward discovery, is organized by scientific disciplines: physical research, materials research, chemical research, mathematics, behavioral sciences, communication sciences, and operations research. Electronics technology (Area 20) does fundamental and applied research on the usefulness of new technological concepts, as well as in the field of manufacturing design. Areas 30, 40, and 50 were the organizational homes of the two exploratory development and two development programs we studied. Bell Labs is a complex organization that must manage and integrate a wide range of disciplines, orientations, and geographical locations.

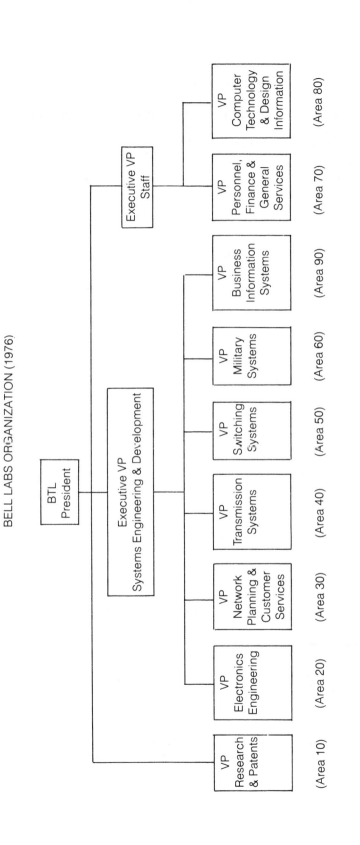

Managing Knowledge Transfer

In addition to managing activities within each of the organization's areas, Bell's executives had to manage the transfer of knowledge through the stages of the technical logic. These stages, discussed in Chapter 3, are shown again in Figure 18.

Our research suggests that a continuum of uncertainties faces R & D organizations, as Figure 19 depicts.

**FIGURE 18
THE TECHNICAL LOGIC OF R & D**

**FIGURE 19
A CONTINUUM OF UNCERTAINTIES FACING R & D PROGRAMS**

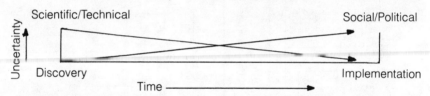

Mansfield et al (1971) stated that R & D "is an activity which is aimed at reducing uncertainty . . . aimed at learning." Each subsequent stage of the technical logic deals with a different mix of uncertainties—generally a decreasing degree of scientific, engineering, and production uncertainty and an increasing degree of market and political uncertainty.

There are implications for management. First, programs should be organized to reduce the appropriate uncertainty. This means defining the research focus to represent a realistic assessment of the existing state of knowledge. As one individual at Bell Labs pointed out, it does not make sense to develop production specifications for something you don't understand. Mounting an engineering effort to solve an uncertain scientific problem is ineffective and inefficient, as was seen in the artificial heart case. A realistic assessment of the research problem also should provide an indication of the time and resources required to resolve the relevent uncertainties before the program can proceed to the next stage.

Second, the organization has to be differentiated to handle the various areas of uncertainty with which it must deal, and be able to shift responsibil-

Organizing the Technical Logic

ity to the sub-unit best equipped for the task. The organizational structure at AT&T reflects the different orientations, or technical sub-logics, which must be part of the R & D process. As the organization reduces the scientific and technical uncertainties, it increasingly turns to the relevant social and political uncertainties. In the Bell System, following resolution of any remaining technological problems in the exploratory development phase, overall responsibility for the program often shifts to Western Electric or to some marketing group, for example. It marks a crossover point where the Bell System organizationally recognizes the shift in task focus. Such crossover points recognize the growing dominance of a new sub-logic, better suited to removing the remaining uncertainty.

Our study also suggests a change in dominant orientation from a scientific to an engineering focus as one moves from discovery to exploratory development and into engineering design; and a shift toward the other pole, implementation, beginning around the development of production processes stage. Organizationally, the development stage appears to reflect a type of cross-over point (Figure 20), which we will describe in more detail, shortly.

A strong desire for appropriate management of both research activities and system development programs, along with a concern for the transfer and use of knowledge, was apparent throughout the Bell Labs. An interest in appli-

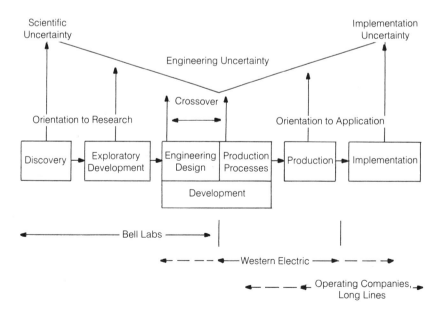

FIGURE 20
THE CROSSOVER

cation in Area 10 seemed to dispel any notion that successful, renowned research scientists ignore application. There also was a positive orientation toward implementation in exploratory development and development programs. In fact, strikingly consistent in the interviews with various people deployed along the technical-logic continuum was their understanding of their respective orientation and role in the organization, as well as the orientations and roles of groups at other positions on the continuum. The descriptions that follow illustrate Bell's approach to organizing for technical tasks and to managing the continuum.

Discovery Activities at Crawford Hill (Stages 1–3)

In the 1920s, Bell Labs established a radio laboratory to conduct research into high-frequency radio transmission. The laboratory remained at its original site until construction of the new Holmdel Laboratory at that location in 1962. The radio laboratory researchers wanted to maintain their independence and avoid absorption by the new, larger organization. They had developed their own unique methods of operating and a climate conducive to a lot of work, free of exacting specifications or red tape. The people at the radio lab fought to maintain their independence and the separate, Crawford Hill facility was built to accommodate them. The reason for non-absorption of the lab seems to have been its success: no one was willing to disturb this productive group.

The Crawford Hill facility, housing the Radio Research Laboratory and the Guided Wave Research Laboratory, employed approximately 80 professionals and a number of support staff; the two lab directors reported to the executive director of the Research and Communication Sciences Division in Area 10. Its work was described as "mainline basic research," the study of the properties of signals, as distinguished from "supporting basic research" such as materials research. Although the primary focus was discovery, the lab became involved in applied projects which often were necessary in support of basic studies; and there was a clear sense of purpose in that the lab recognized it was only a part of a larger system whose business was communication.

> Our job is to find out what nature makes possible. It's someone else's problem to see whether it becomes a part of the Bell System. For example, Long Lines can now handle 500,000 conversations, but a multi-beam satellite system could handle 50 million. It may not be economical, but nature makes it possible to put a lot of circuits up there.
>
> ***
>
> We'll look at extreme cases and idealize situations to build math models. We hit the high points. We fix a lot of variables and try to choose the ones we think are most important. The development groups worry about Chicago and Los Angeles.

Organizing the Technical Logic 105

Nature leads research in some areas. Sometimes we have to build devices to continue our investigation and experiments. To explore nature in its physical form you often have to assemble a hand-made research model.

***

We have a mission. We know what business we're in and use it every day in making decisions.

Crawford Hill was a small, flexible, and informal organization that strived to maintain a campus atmosphere. In fact, throughout Bell Labs, this atmosphere was encouraged by a variety of mechanisms such as complete technical libraries at each laboratory, relaxed dress codes, publication support, and seminars. At Crawford Hill, there were even students working on PhD theses. One research scientist commented that he was delighted to have the students there—they asked a lot of questions and forced him to keep up to date. A perceived advantage that Crawford Hill had over universities, however, was a norm of cooperation.

This place is small and informal and the lines of authority are on a personal basis. If you want to drill a hole in the wall, you drill it. As for the shop, we authorize it a lump sum for three-month periods and everybody uses it. It's a complete service shop—you get done whatever you need. We've reduced a lot of barriers here. We're very jealous of our situation. But you couldn't run Long Lines this way; it would be chaos!

***

The lab is very loosely structured in terms of assignments. The way it functions is dynamic; it's always changing . . . We maintain a close relationship between groups and labs. People are intermingled throughout the building. There is a conscious effort to mix people in the building. There is an appreciable advantage to keeping people rubbing elbows.

***

We try to put people into slots they're fitted for. Some like basic research and write a paper each year for the Astrophysical Society. Others like to see things happen. We try to hire for the job we need done. You have to fit people and jobs.

***

Here we help each other, we don't operate in a zero-sum game. In a university, it is a question of survival; here everybody has tenure.

The Radio Research Lab

The Radio Research Lab provides a good example of the type of activities and operating methods found at Crawford Hill. The lab director pointed out that the lab did a mix of applied and fundamental research which also appeared to be mirrored in its departmental structure.

The point-to-point radio station group was studying alternatives to laying cables underground in metropolitan areas, such as using highly focused signals transmitted short distances by metallic reflectors on roof-tops. Work on this system had been stopped because the technology to build it did not exist; the operating companies could not install it; and there was not enough communications traffic between buildings to utilize its capacity. The advanced satellite group's primary focus was the development of systems concepts; they were working with AT&T marketing and engineering to identify the future needs of the Bell System. The radio propagation studies group had been studying the effect of rain on radio transmission for fifteen years, as well as pursuing research on multi-satellite systems.

Probably the department that showed just how "basic" the research at Crawford Hill could get was radio astronomy and devices. Arno Penzias, department head, and his co-workers had made some important discoveries, such as the existence of carbon monoxide in space. Their most significant contribution was the discovery of the cosmic background noise of the universe, which, according to the lab director, was a "monumental discovery that could get them a Nobel Prize."[1] Penzias stated that he was oriented both to research and to application and that sometimes the critical interface between the two was not always between people, but sometimes inside people like him:

> I suffer from schizophrenia. I like to do things that have applicability, and things that may seem totally useless. From the phone company's point of view, we want to be a source of knowledge as well as a sink.

At one extreme this group was interested in the formation of the universe and such issues as the nuclear composition of interstellar molecules. As Penzias explained:

> Radio astronomy began at Bell Labs. This group is concerned with what's out there in the universe, what's going on, and how it all got started. It's basic science.

However, his group also had responsibility for developing electronic devices, like integrated circuits, for use at the high frequencies at which they were working, since such devices could not be bought; and for the design and construction of the antenna to be used in satellite communication rain studies. Although Penzias and one other person could have devoted their full time to astronomy he chose to split his time between company projects and astronomy and to maintain links both within the company and with the external scientific community.

Funding and Project Selection

The primary control over Crawford Hill was the size of the organization. The lab had a personnel authorization which pretty well determined the size of the budget. However, each year Crawford Hill submitted a proposal for the upcoming year's work.

---

[1] Arno Penzias and Robert Wilson did, in fact, win the Nobel Prize for Physics in 1978 for their discovery of the cosmic background noise which supports the "big-bang" theory for the formation of the universe.

Organizing the Technical Logic 107

The proposal, approximately ten single-spaced pages, was reviewed up through Bell Labs and AT&T and partially was the basis for the labs' multi-million dollar budget. In addition to the formal proposal, progress reviews with people in AT&T, Western Electric, and other parts of Bell Labs, as well as visits to the lab by groups from these organizations, communicated the labs' work. According to the Guided Wave Lab director:

> These verbal interchanges provide the basis for the proposal. Paper is not a substitute for real understanding of what's happening. People on both sides of the piece of paper know what's going on in the lab.

Although radio and guided wave research were the labs' primary missions, they were basically "umbrellas." The lab directors did not feel overly constrained by their missions and had considerable autonomy and latitude to pursue areas of interest. They did not require specific authority to pursue interesting ideas or start a research project as long as they remained within budget.

Project selection was also largely an informal process. Most projects started with an idea generated by a member of the technical staff who discussed it with his department head. If it involved a fairly large commitment of time and money, the lab director would participate in the talks. Occasionally, management might suggest a possible research topic and, through a series of informal discussion, gain agreement. In all cases, the process relied on the mutual judgment and opinions of departmental heads and lab directors for starting and stopping projects. The Radio Research Lab director indicated that pressure was not part of the process:

> I have never been told to do anything, nor have I ever told anyone else to do something.

Communicating Results

The results of the work at Crawford Hill were communicated in many forms: word of mouth, visits, quarterly progress reports, internal memoranda, seminars, and work reviews. But one of the primary modes was publication. The researchers at Crawford Hill were prolific contributors to scientific journals, according to one of the directors:

> People here are writing all the time. Some have written books. Some are done on company time with secretarial assistance, some aren't—we're not sticky about that. People like this work so hard you don't worry about them taking advantage of you; you worry more about their health.

Another measure of output was patents, but success at Crawford Hill was not measured simply by patents, publications, and professional awards. A frequent comment heard was that there were many ways to display knowledge. Another important criterion was the application and implementation of an idea. Commitment to an idea and responsibility for it did not end until it was accepted for exploratory development—and sometimes this meant even building a prototype system!

> The big success of Crawford Hill is TD2, which is on top of all telephone buildings in the country. It is the backbone of the transmission system and was invented here at Crawford Hill.
>
> ***
>
> We have a lot of tradition in this place. We're activists; we want to see something happen. We'll build the system on roof-tops in New York, if necessary. It's hard for people to argue that you can't develop a system if you do it and show it to them. We'll drop an idea when we think it is getting a fair shake.
>
> ***
>
> We tried to get people interested in developing this (DR18), but no one wanted it. It was our concept, so we built a working model. We also had to convince ourselves that it worked. We do whatever is necessary. It has been a big success—maybe not economically yet, but environmentally and technically. It works like a charm.
>
> ***
>
> There are some people in the Bell System who want to maintain the *status quo*. That's why we build prototype systems and shove them down their throats.
>
> ***
>
> We have to get above a threshold of what other people in the Bell System perceive to be useful. We have to put together a skeleton version so people can see that it is real—physically real.

Exploratory Development (Stage 4)

Our review above of Bell Lab's activities at the discovery stages makes it is easy to see how knowledge generated in places like Crawford Hill becomes available to the rest of Bell Labs and how ideas for systems find their way into exploratory development. The high level of interaction and communication throughout Bell Labs permits researchers to display their wares and for other parts of the organization to express their needs. This continual push of findings and ideas from researchers and the pull for ideas and technology from development groups and customers like Long Lines, for example, is fostered and supported by the research and engineering executives at all levels.

In Chapter 3, we described an exploratory development program, millimeter waveguide, in detail. In that case, we saw an emphasis on designing a high quality, complex, working system; a concern for integrating the program with groups at subsequent stages of the technical logic; a recognition of the necessity to loop back to a prior stage to pick up knowledge and technology as well as the willingness to do so; and the review mechanisms that integrated the program into overall system planning. Rather than repeat that detail, we refer the reader to the discussion of millimeter waveguide as one example of an exploratory development program.

The High Capacity Mobile Telephone System also had completed explo-

Organizing the Technical Logic

ratory development at the time of our study. In Chapter 2 we examined how this program managed its political environment. We also looked at its technical management history and discovered organizational processes similar to those that contributed to millimeter waveguide's technical success.

First, there was the emphasis on coordination that tied the program into the plans of the Bell System. A combination of the FCC's receptivity to increasing radio spectrum allocations and AT&T's decision to scale down its military effort and shift people to other activities utilizing similar frequencies led to the formation of the Mobile Communication Laboratory in 1970. The lab was to convert high frequency research into workable commercial systems. At this same time basic radio researchers were ending their studies on this particular frequency band and, in a series of meetings, passed along their experience to the lab engineers.

Next, came meticulous attention to designing a system and understanding how it would function in its ultimate physical environment-urban areas. Researchers had dealt with statistical models based on carefully controlled test data gathered in a "clean" environment by sophisticated, precision antennas. Lab personnel experimented with putting transmitting antennas on high buildings, in heavy industrial zones, in suburban areas, and prominent terrain zones. New technology and new analytic methods also were developed during this period to support the development of a working test system which was followed by intensive field experimentation.

Finally, in 1974 when the FCC decided to allocate spectrum space for cellular systems, AT&T made the commitment to develop the system, and program leadership formally switched from Bell Labs to AT&T Marketing. AT&T Marketing continually had monitored the mobile communication market and assumed responsibility for system development as the customers were outside the Bell System.

These exploratory development programs went through a change in operating methods as they gradually defined the systems, resolved technical uncertainties and moved toward the development stage. Comments from Mobile Lab personnel indicate this progression.

> When we formed this group initially, we had an idea that the frequency would be around 900 MHz, but nothing really was certain. The FCC said they were interested in improving mobile service and were casting around for ideas. We were really reading tea leaves in those days.
>
> ***
>
> It started in an unstructured way. Although we laid out a five year plan and tried to indicate manpower and dollar requirements, we really didn't set any dates or specific timetables. We identified the major activities necessary to bring this thing to life. The only real firm date was December 1971 when our report was due.
>
> ***
>
> When we were working on the initial report, milestones were a year apart and jobs not well defined. Progress was made on an annual basis and progress re-

views were held monthly or quarterly. In 1974, we snapped into a highly structured exercise. We've switched to almost daily control since we've become more tightly coupled with Western Electric and outside suppliers.

Once major technical issues were resolved and the Bell System was convinced that a new system could and should be manufactured, a program moved into stage five development.

Development (Stage 5)

In the Bell System, the development stage is a clear example of the internal juncture of two sub-logics, and Bell uses an interesting structural response for managing this interface-branch laboratories. A branch lab is a decentralized unit of Bell Laboratories located in a Western Electric production facility responsible for the engineering design necessary to prepare a system for production.

Branch labs provided a vehicle for moving new technology from Bell Labs into the production processes of Western Electric. Although the new technology had been tested under controlled conditions in Bell Labs, it had not necessarily been tried in large-scale production conditions using less-skilled workers. The branch lab organization also provided prompt solutions to manufacturing problems arising from design specifications, and permitted modifications to be made right at the manufacturing locations.

Jack Morton (1964, 1967, 1971), who was a vice president of the Bell Laboratories, explained this administrative mechanism by what he referred to as barriers and bonds theory; or what we might call managing logical boundaries. Communication is extremely critical to the functioning of a knowledge development organization and Morton pointed out four factors which contributed to, or hindered, task communication. These were *specialized-language, space, organization structure*, and *motivation*. To ensure the full flow of information, Morton provided a relatively simple heuristic:

> Using these four communication factors, we can deliberately build into the system certain bonds and barriers that will either inhibit or encourage the flow of information. These bonds and barriers are used in complementary fashion; whenever we have a spatial barrier we try to have an organizational bond and vice versa. Two barriers never exist together lest information flow be impeded and two bonds never exist together lest one specialist group dominate the other. (1967; 33).

Morton, in reviewing the history of the Bell system, observed that, until the end of World War II, two principal barriers impeded communication between Bell Labs and Western Electric—organizational and spatial barriers. The spatial barrier was removed by physically housing Bell Labs' development design people in Western Electric facilities, thus introducing branch labs. Barrier and bond theory seems to be a fine example of an heuristic for

Organizing the Technical Logic

maintaining positive tension at an important internal boundary; and branch labs, an example of the application of this theory.

The Branch Lab at the Merrimack Valley Works

The purpose of the branch lab was to link designer, user, and manufacturer together so that nothing was overlooked and the lab personnel had a very specific, short-term focus to their work.

> We can't design equipment without knowledge of the user, the operating company and its environment. We also have to look at the manufacturing process to make sure the thing we design can be made. You can't develop systems or change them without understanding all parts of the process.
>
> ***
>
> We don't study something here for 10 years. Our job is implementation and that's different than the jobs at Murray Hill or Holmdel where I was previously. We should be looking right down the sights of a rifle and not shotgunning. We have to be specific and 90 percent of our effort should be on a product that will be manufactured in a year or two. Holmdel can be more long-term oriented and diffuse. We are more sensitive to the needs of Western Electric and the customer. This is our role—somebody has to do it.
>
> ***
>
> I get my kicks seeing what we develop—immediate results, not a detached view. The world needs some paper writers, but I'm not one of them.

Linking designer, user, and manufacturer meant the exchange of a lot of information. The branch lab, as an organizational part of Bell Labs, had ready access to the relevant basic researchers and design engineers. In addition to information about operating companies' problems and environments obtained through the normal range of planning and review meetings, branch lab people got it "directly" by putting their own people in the field environment temporarily and by borrowing operating company personnel with specific experience or expertise. Manufacturing information was readily attainable since the branch labs were co-located with Western Electric. Co-location was perceived beneficial by Western Electric's product engineers, works management, and by Bell's designers as the following comments indicate.

> Having Bell Labs people right out there on the floor when you have problems is a lot nicer than talking on the phone to them in New Jersey.
>
> Nothing could be accomplished without this attitude in such a large system. Being located so close to Western Electric means our output has to contribute to the people across the hall. A two-way flow of information is important. The job's much tougher if Western Electric isn't part of the design process. Some remotely designed systems suffered because they didn't have the day-to-day interaction with Western Electric engineers.
>
> ***
>
> Bell Labs does the model building—producing one or two models. But it's a much different situation when you get into production and the tyranny of numbers. No one could be perceptive enough to see all the problems you could pos-

sibly run into. That's why we have different groups looking at a program from different perspectives as early as possible. There is no way we could do it without them.

Looking back on the histories of waveguide and mobile telephone, one is impressed by the time required to transform an idea into a system ready for stage five development. Then, as the system enters into development and manufacturing, one also is struck by the variety and number of groups that must be coordinated to produce a complex system. Time frames shorten, interdependency increases, and interactions become more frequent. The two development programs in our study sample, the D4 terminal and the No. 4 Electronic Switching System provided an opportunity to observe how a system that may have had its beginnings in a place like the Radio Research Lab finally finds its way into production at Western Electric.

The D4 Program at the Merrimack Valley Works

D4 is a terminal in a telephone office into which telephone lines are connected. It is an evolutionary product designed to provide greater capacity and to transmit data faster than the system it replaces. In preparing the development proposal, Bell engineers spent a couple of months making cost and technical analyses, and spent days talking with the telephone companies about desirable features and what they would be willing to pay for these features. After the analysis and exchange of ideas, the proposal was written. D4 was funded for development starting in 1974.

By April 1976 four prototypes of the D4 terminal had been built by Western Electric labor in the terminal department of Bell's branch lab at the Merrimack Valley Works. Parts were being supplied by manufacturers whom the Western Electric purchasing organization had selected as suppliers. Two of these prototypes were sent to Holmdel for experimentation and study; one went to Indian Hill, for exploratory studies on linking electronic switching systems with transmission systems, and one went to the Western Electric works in North Carolina that ultimately would produce D4. When D4 went into production at the end of 1976, program personnel would have had experience with the suppliers and their products, and Western Electric and Bell Labs engineers would have had experience building, testing, and debugging the terminal.

Although the terminal was scheduled to go into production shortly, there was still work ahead for the Bell Labs engineers:

> After production we will need a good round of cost reduction. We will also have to follow the system into the field and be intimately involved with it for a while. We'll be working on D4 through 1980.

And even while production had not started and cost reduction efforts spanning the next few years were being planned, D5 was already in mind.

Organizing the Technical Logic 113

You're always thinking about a better way. The managerial problem is knowing when to bound the existing problem and say let's make this product. The exploratory development for its replacement starts with the changes you couldn't or wouldn't make in D4, for example. Most of the poor products from Bell Labs have resulted from not knowing when to stop making changes and produce them.

The information transfer sequence at this stage is shown in Figure 21. Although the diagram depicts a sequential operation, the people involved strongly emphasized that in reality it was more simultaneous in nature.

FIGURE 21
TRANSFER OF DESIGN INFORMATION

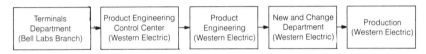

Bell Lab designers sent laboratory design information specifications (LDIs) to the Product Engineering Control Center (PECC) at Western Electric. The PECC, formally designated as the design-production interface, translated the LDIs into manufacturing specifications. The PECC at Merrimack Valley controlled system standards and specifications for all transmission products made there as well as in Kansas and North Carolina plants. It codified, standardized, and disseminated design information. It ensured technical compatibility between the design and the needs of diverse groups. As indicated in Figure 22, the users of this data were both the manufacturing and field operations.

PECC personnel worked with Bell Lab's engineers to influence the manufacturability of the design. Early in the program, they reviewed design features like frame connections for compatibility with existing central office equipment. Although approval of a Bell Labs development proposal was the first official notification of a new product, word spread informally and information about specifications often was picked up early; some documentation would be completed before the arrival of LDIs. For example, even before LDIs on D4 were completed, Western Electric's production engineers had contributed their knowledge as a design input.

Product engineering looked after physical facilities and equipment. Product engineers secured the capital equipment and floor space in the plant required for production, and arranged for tooling, test equipment, and testing of manufactured devices. For D4, they also coordinated the manufacture of more than 600 different components at numerous Western Electric and outside supplier locations—260 of which never had been manufactured before.

The new and change department secured long-lead-time items, ordered all

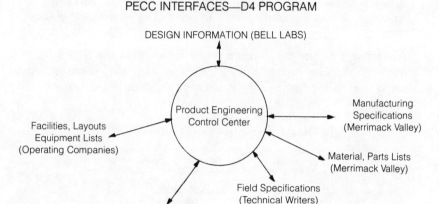

FIGURE 22
PECC INTERFACES—D4 PROGRAM

parts, scheduled and monitored parts flow, and transmitted design information (including changes) to the production shops. It handled these functions for six major product lines of new design, and for 1300−1500 changes to components. In addition to ensuring the availability of parts for D4, this department monitored the overall production schedule, "pushing and chasing" engineers to meet the availability date and to work with the operating companies on installation procedures. It also sequenced production between the old system and D4 to stabilize overall plant employment. Because D4 would not be used by some telephone companies, production of the old system had to be scaled down carefully and balanced with the start-up of D4 to avoid layoffs or equipment shortages.

Bell Labs designers held meetings every other week with Western Electric engineers on the status of manufacturing arrangements. At this stage of the process, the terminals department head called and chaired the meetings, but Western Electric would assume this responsibility once production started.

The Number 4 Electronic Switching System (No. 4 ESS) Program

No. 4 ESS was the largest and most expensive development project ever undertaken by the Bell System. More than a dozen Bell Labs, Western Electric, and AT&T units took part in developing the system, first put into service in Chicago in 1976. It is the world's highest capacity switching system, capable of routing approximately 550,000 calls per hour or four times the volume of any existing switch in the Bell network. It was developed and manufactured in response to the growing number of interstate calls, anticipated to be 40 billion by the year 2000—a ten-fold increase from 1975.

No. 4 ESS did not appear out of the blue. It was the offspring of a long

Organizing the Technical Logic

history of technological development. The first electronic switching machines combined electromechanical and computer technologies. A digital computer controlled the switching system using stored, programmable routines rather than hard-wired connections. Although No. 4 was the newest in the family, it was the first used for long-distance switching and did not replace other ESS machines. The system it replaced, 4A crossbar, had been Bell's first application of photo-transistors. Introduced in 1952, it permitted nationwide direct dialing. In the late 1960s a computer was linked to 4A crossbar, increasing capacity, but the system limits were being approached. In addition to building on past developments, No. 4 ESS incorporated new features. Its use of pulse code modulation made it the first digital switching system; its special control unit provided extremely rapid switching, monitored its own operation, and provided automatic network management capability; and it used time-division switching.[2]

The origins of No. 4 ESS can be traced back to the early 1950s when Earle Vaughn, director of switching research at Murray Hill, built a solid state, digital, time-division switch. Another Bell Labs' group built a solid state, space-division switch which experimental data indicated could be developed more economically. This system became No. 1 ESS. Vaughn transferred from the research department to the Indian Hill Lab outside Chicago in 1962 to become director of systems and software for No. 1 ESS. He left Murray Hill, however, with the understanding he eventually could pursue a time-division switching system.

From 1968 through 1970, when the No. 4 ESS program was funded for development, many comparative studies of competing technologies produced a serious tug-of-war within Bell. One faction advocated the old technology because it knew a viable system could be built with it, while another felt the old technology had been milked dry and it was time for new ideas. In 1970, the issue was resolved with the decision to pursue No. 4 ESS with its new technology. The executive director of toll switching systems at Indian Hill explained that even with their tremendous emphasis on collecting data the decision was far from routine:

> The decision was based on judgment. Looking back, few facts really existed. We knew we could do it technically because we had done it in the lab. The real question was whether we could manufacture it and the risks were around meeting cost and time schedules.

Organizational Decisions

The Bell Labs facility in Columbus, Ohio, had responsibility for "toll" (long distance) switches, of which No. 4 ESS was one, while Indian Hill developed "local" electronic switches. Columbus was home of the 4A

---

[2] In time-division switching, many calls are sent simultaneously over the same physical path separated by fractions of one millionth of a second intervals. In the older method, space-division switching, it was necessary to have a separate physical path for each conversation.

crossbar family of products and also the locus of the technological orientation toward electromechanical devices and space-division switching. Indian Hill was oriented to electronic, computer-controlled systems and time-division switching.

No. 4 ESS was located at Indian Hill because its technological focus was the most appropriate. Only an all-electronic machine could achieve desired switch speeds. However, locating the 4 ESS Lab at Indian Hill under Earle Vaughn, but reporting organizationally to Columbus, created some tension.

A second decision affected the relationship between switching and transmission systems area. In the 1950s, switching and transmission systems were part of the same organization. These were split in the early 1960s with the advent of complex, analog computer-controlled switches (ESS machines). During the following decade, the two remained organizationally and technologically separate. Transmission systems increasingly became digital while switches remained analog like most of the Bell network. Engineers at Holmdel, for example, had experience designing analog-to-digital interfaces for their transmission equipment. No 4 ESS had to be digital in order to utilize time-division technology, and Indian Hill needed to import this interface design experience.

In response, a group composed of switching and transmission engineers was set up with a simple assignment—design the analog-to-digital converter for 4 ESS. When these people started focusing on the transmission-switch interface, they discovered ways to achieve significant cost and space savings by tying the transmission line directly to the switch. It would be possible to go to a completely digital toll network. The director of the No. 4 ESS Lab observed:

> No. 4 ESS is beginning to change the traditional differences between switching and transmission. It's beginning to overcome the differences of separate organizations 800 miles apart which contain people of varied backgrounds.

Another Bell Labs executive commented:

> Innovation in this area had fallen into the organizational "crack" between transmission and switching. The cracks, or where you perceive them to be, keep moving around and you always have to consider them.

Another, more subtle, change which had its roots in the previous technological decisions was taking place. Computer software technology was becoming dominant. Software designers and software systems were influencing hardware design rather than the other way around as had been customary. This change was being felt all the way from craft level employees in operating companies, who had to maintain the system, through designers at Bell Labs. It required significant changes in maintenance procedures, training and hiring policies as one executive explained:

> We used to hire electrical engineers with circuit design knowledge. Now we

lean toward computer scientists or EEs with computer science experience. We need this blend.

The organizational interfaces were as involved and complicated as the technical ones. No. 4 ESS was owned jointly by the operating companies which arranged all interfaces and space requirements with local offices, and Long Lines which operated and maintained the toll switch. The primary control points for the program responsibility rested with Bell Labs at Indian Hill, while the Northern Illinois Works (NIW) was the headquarters for Western Electric's program management. Figure 23 presents a simplified overview of the complexity of organizational involvement in the No. 4 ESS program.

A wide variety of administrative mechanisms was used to link all the involved organizations. There was the standard array of inter-company committees, priority committees, change committees, an installation planning committee, and a committee focusing on initial operating issues. In addition to these common mechanisms, we came across three integrative devices not seen at Merrimack Valley.

AT&T Engineering. This organization played an early role in initially matching a system's capability with the needs of the user. A notification of design intent was sent by AT&T to operating companies outlining a proposed system's capabilities. A preliminary letter without much technical detail was issued for planning. It presented facts on the increased capacity and additional service features beyond that of existing equipment. A second letter spelled out detailed differences between systems.

Project Management. Both Bell Labs and Western Electric organized specific project management functions to coordinate the increased organizational interdependency of the No. 4 ESS program. These were levels of coordination beyond those previously existing in development programs.

(a) *Bell Labs*. The project manager at Indian Hill believed he was the only person with that title and function in Bell Labs. He coordinated activities with all design labs and manufacturing plants to ensure that "the wheel was not re-invented at each location" and to minimize the number of new designs. Although technical direction was usually one of the responsibilities of Western Electric's Product Engineering Control Center (PECC), Bell Labs provided it in the early development phase and PECC engineers were assigned to Bell Labs project manager in 1970. They eventually returned to Western Electric to contribute to its planning effort.

(b) *Western Electric*. No. 4 ESS was the first "officially chartered" project management activity in Western Electric. A temporary management system, ceasing after product introduction, supplemented and integrated traditional, functional groups. Project managers were chartered by Western Electric vice-presidents to cut across traditional lines of authority at all levels in order to manage the system introduction process.

Computer Systems. The SPIDER (Shared Production Interactive Data

FIGURE 23
BELL LABS AND WESTERN ELECTRIC ORGANIZATIONS INVOLVED IN THE NO. 4 ESS PROGRAM

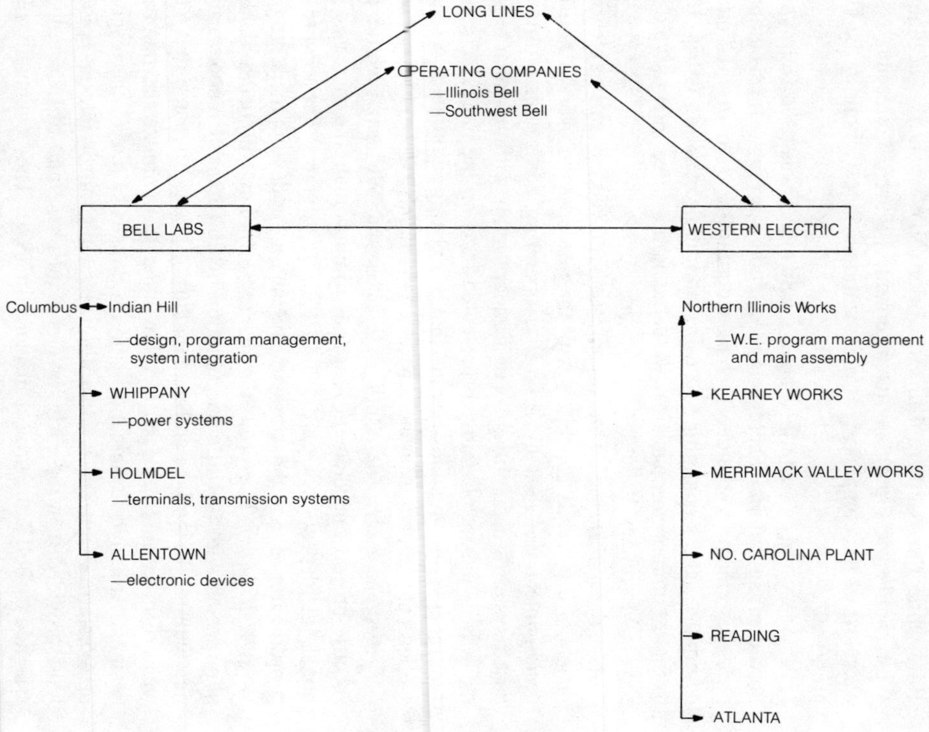

```
                    BELL LABS  ⇅  WESTERN ELECTRIC
                         ↕↕                ↕↕
                      LONG LINES
                   OPERATING COMPANIES
                      —Illinois Bell
                      —Southwest Bell
```

Bell Labs:
- Columbus ↕ Indian Hill — design, program management, system integration
- WHIPPANY — power systems
- HOLMDEL — terminals, transmission systems
- ALLENTOWN — electronic devices

Western Electric:
- Northern Illinois Works — W.E. program management and main assembly
- KEARNEY WORKS
- MERRIMACK VALLEY WORKS
- NO. CAROLINA PLANT
- READING
- ATLANTA

Organizing the Technical Logic 119

Base for Error Reduction) computer system contained common design data base which replaced a manual system for controlling engineering change orders. This system bridged the gap between the design ideal and manufacturing reality. Complete circuit design information is stored in a data base at Indian Hill. The Northern Illinois Works, using this data base, automatically prepares circuit layouts and test procedures used to manufacture equipment and to test its manufactured components and systems.

SPIDER provided the means for evaluating proposed changes; for generating integrated sets of documents reflecting software, operating, production, and testing changes; and for checking the compatibility of new specifications with the requirements of affected systems.

Implementation

Engineers and craftsmen from Long Lines and the operating companies were sent to Indian Hill in 1972 to start learning the system. Since no field testing was scheduled and these individuals needed to be familiar with the equipment, they learned test and repair procedures at Indian Hill. In August 1974, Western Electric delivered the Chicago system to Indian Hill for testing, and months of intensive hardware checks followed. Meanwhile, the control programs were being tested on a computer simulating the hardware system. In January 1975, the machine was installed in Chicago and the software and hardware systems were linked for the first time.

As part of his responsibility for maintaining Bell Lab's design standards while meeting Illinois Bell's and Long Lines' operating objectives, the Bell project manager supervised the process of "putting all the pieces together." His Chicago test group of 9 engineers (including the two people temporarily assigned from Long Lines) spent a full year on-site-testing and debugging programs. Part of this year was also spent linking the Chicago equipment into the lab system at Indian Hill so that it could be monitored remotely and assistance in analyzing problems could be given to field personnel.

In January 1976, operations began on schedule. Start-up was designed as a gradual process to maximize learning about man-machine interfaces. By April 1976, the entire system was operating and the test group was looking forward to similar, but shorter, six-month installation in Kansas City. The test group's involvement in start-ups would diminish gradually as Western Electric, the operating companies, and Long Lines gained experience in producing and installing No. 4 ESS.

Linking Task and Organization

Bell Labs has been quite successful in generating knowledge and producing technology. Without claiming to be the only reason for this success, we believe the organizational arrangements of the type described in this chapter make a significant contribution to that success. Not only were specialist groups in place to resolve problems specifically related to the stages of the technical logic, but the importance of linking these groups to ensure the transfer of relevant information and skills was emphasized. It was clear to us from interviews with Bell's executives that organization, as well as technical specialties, mattered to them. Knowledge and organization were necessary components in developing and producing communication systems. The link between Bell's task and its organization is reinforced by the concept of the Bell System as one system with physical and organizational components.

The physical component is the telephone network itself and programs had to be justified in terms of the total operating communication system. All parties—researchers, developers, and managers—were fully attuned to this governing image of a single system. Bill Warters, the man who probably had the largest and biggest stake in pressing on with the millimeter waveguide development, said it all in commenting that the most recent delay was "personally disappointing but not disappointing from a Bell System standpoint. If it's not economical, it is not right".

The "one-system" concept emphasized technical and market feasibility of each project before funding full-scale development. There also was emphasis on acquiring and using intimate knowledge of all segments of the environment bounding the system under development. The end result would be a product designed to fit into and enhance the existing network.

New technology created waves lapping Western Electric, often creating difficult times and tension. The more successful Bell Labs was in creating technology, the fewer people were required by Western Electric which relied on Bell Labs for its product specifications. As Western Electric added a person to build electronic devices to increase telephone line capacity, it might lay-off three building cables. Both the development of tightly integrated circuit-board components, requiring mounting of fewer elements, and the convergence of telephone technology with computer technology, bringing competition by many qualified component suppliers, could significantly affect Western Electric's direct labor requirements. Organizations must adapt to these changes. As one Bell Labs executive puts it:

> Our organizational arrangements are dictated by the things we do and are dependent on the technology involved. You have to build your arrangements around the physical phenomena. The organization has to mirror this.

The one-system concept provided managers throughout Bell with a powerful image that forced them to recognize such interdependencies and to understand actions aimed at balancing the organizational system. The manifestation of this type of thinking was seen in the array of structures, systems, and interpersonal processes used to face-off organizational units to a particular stage of knowledge development and, simultaneously, to tie them together.

The two components ensured effective information transfer, that is organizational messages as well as physical telephone messages. The output of this over-arching one-system concept was a design, either product or organization (Figure 24), intended to improve the existing system.

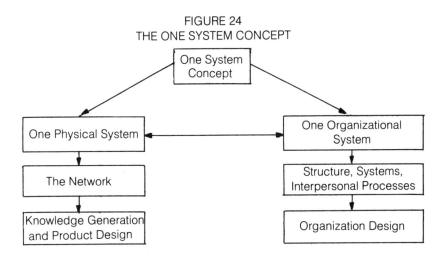

FIGURE 24
THE ONE SYSTEM CONCEPT

A strength of the Bell system is the degree to which it distinguishes differing task situations and matches its administrative processes to fit them. The mirror metaphor and the one-system image were in active use in the Bell system. Both seemed to function as heuristic aids for fitting the organization to its task and environment and as criteria for managerial action.

What can the Bell experience teach other R & D organizations like NIH? The Bell System is more highly differentiated and integrated than NIH. It simply has more organizational units focused on the various stages of the innovation process and on linking them together than does NIH. It is organized specifically to move knowledge and equipment through this pipeline. NIH is primarily a research institution that is not organized for development and implementation processes, although we have seen that it becomes involved in these areas, often with less than positive results. If the political system wants NIH to get fully into the development business, it will have to give thought to ways of organizing to achieve this purpose. And it will have to be more realistic in its expectations. We have seen that the process of moving knowledge from discovery to implementation is complex and time-consuming even when you are organized to do it. The Bell experience illustrates the important relationship between technology generation and organization. The successful R & D organization will be conscious of the different orientations necessarily embodied in it and will organize to reduce internal barriers to promote transitions in technical stages.

Chapter 7

Adaptation to Changes in the Technical and Political Environments

R & D executives must manage complex sets of technical and political logics to adapt a program to its multiple environments. An effective organization attracts the resources required to maintain itself and achieve its goals: employees, funds, capital equipment, as well as social and political support. To remain effective it must continually relate to the evaluating and supporting elements of its environment and moderate their concerns with the program's need to generate knowledge. An efficient organization will actively manage and integrate its internal technical sub-logics to keep programs on-track and synchronized with its requirements and constraints. A program or organization that cannot reconcile a conflict between a primary technical focus on discovery, for example, and outside pressures to produce quickly practical results will have a difficult time accomplishing either. We refer to a balanced combination of efficiency and effectiveness as program integrity. It is virtually axiomatic that for a program to succeed it must integrate its political and technical logics. Otherwise, a loss of program integrity develops.

The NIH program that probably had the greatest difficulty in this regard was sickle cell. At the risk of oversimplification, we saw this program using a technical logic suited for discovery which fit the state of relevant knowledge. Its political logic, however, reflected a noble effort to be responsive to the sudden and volatile concerns of major constituent groups. No effective action was taken to reduce public expectations to match the hard facts of limited knowledge and no true therapies. A gap developed between the political logic and task logic that was widened further by divided authority. No superordinate organizational logic closed this gap. A single triggering event sending a wave of disillusionment through the constituent groups discredited the program for a period of time. Perhaps these events were unavoidable but, regardless, they can provide a useful example for future guidance.

The Artificial Heart Program provided another useful lesson. The program maintained an extreme consistency between its political and technical logics. Its technical logic was to assume a development effort employing systems-thinking and a master plan. Its political logic was to lobby in Congress for a fully implantable heart, and ignore criticism. The choice gave the program a misdirected internal consistency even as evidence was building of a lack of fit both with its technical realities and with part of its political environment, namely, the Office of the Director of NIH.

In all the NIH and Bell programs that sustained a condition of program integrity, a simple rule of thumb seems to have helped tie together technical and political logics: *The knowledge of the physical phenomena should govern the choice and sequencing of technical activities, but social/political realities should control the timing of the effort, and both will influence the scale.* The use of such a heuristic was most obvious in the Waveguide Program. This program twice was put on hold in spite of continuing technical success because of the realities of market demand and competing technologies. When these same factors justified restarting, the program renewed activity at the stage congruent with the technical realities. The third clause in our rule, concerning the scale of the effort, can be illustrated by the SIDS Program. A high level of funding offered by Congress early in the program was resisted by program leadership as conducive only to poor and confusing research. This Congressional interest, however, did convey to the scientific community a sense of urgency about the problem, and definitely influenced, but did not totally control, the pace and scale of the effort.

We discussed the relationship between Bell's organizational and technical logics in the last chapter and saw the strength and benefits of institutionalizing this relationship. We have no example from our sample of programs of a similar relationship with an organization's political logic: an institutionalized version of what Dr. Hasselmeyer personally accomplished in the SIDS Program, although the High Capacity Mobile Telephone System Program probably came the closest. We do, however, have an example from NIH, the Genetics Program, of a situation in which all logics were modified over time to adapt the program to changes in both its technical and political environments.

The fact that an organization has adapted once to its technical and political environment may influence its future adaptiveness, but does not guarantee its ability to perform similarly in the future. Only by making the dubious assumption that the environment will remain stable does current adaptation imply long-term adaptability. Recent experience does not seem to justify the assumption. An historic perspective is needed to understand adaptability. By analyzing a number of significant changes and organizational responses one can attempt to identify the factors associated with the movement from one pattern of adaptation to another. Shifts in the Genetics Program that allowed it to remain adapted to its multiple environments can be described and explained using our model. The evaluation of this program also illustrates the dynamic, interactive nature of technical, political, and organizational logics.

FIGURE 25
CHAPTER 7 FOCUS

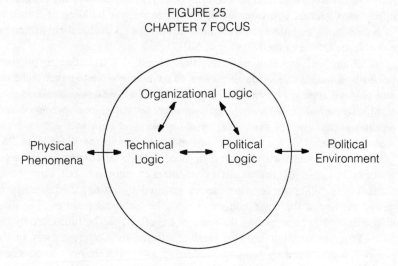

Origin of Genetics Research Activity at NIH

The origin of organized genetics research activity at NIH was notably non-controversial. Private discussions among some prominent scientists and the director of NIH led to the idea of a general, basic science program. It later was discussed in a meeting attended by an important congressional figure, Senator Hill, who viewed the idea favorably. The proposal was brought to the personal attention of Congressman Fogarty whose subcommittee already had completed hearings on NIH appropriations for fiscal year 1959. Congress, in response to the belated request, further appropriated 20 million dollars for fundamental research projects and 10 million dollars for training grants in the basic medical and biological sciences. This was the origin of the Division of General Medical Sciences which became the National Institute of General Medical Sciences (NIGMS) in 1963.[1] One of the major reasons for proposing the budget appropriation in 1958 was the increasing scientific activity in genetics research. Although the funding of NIGMS was not directly attributable to the Watson and Crick discovery of DNA structure, the state of scientific knowledge had progressed far enough for such support.

The institute's areas of research are genetics, and cellular and molecular biology. The biological processes on which it focuses underlay all diseases

---

[1] We will use the abbreviation "NIGMS" to refer to this organization even when talking about it prior to 1963.

and medical disciplines. As a result, the projects and applications referred to NIGMS have always been those either concerning two or more institutes or of long-term basic relevance to biomedical knowledge-building. NIGMS is one of the few institutes in NIH not tied directly either to specific diseases or an organ system.

The institute grew considerably, although perhaps not as dramatically as some others like the Cancer Institute. During the period 1958 to 1963, funding expanded from $30 million to $143 million but, thereafter, growth was marginal, especially considering inflation. In 1975, funding reached $187 million. Genetics research generally has received between a quarter and a third of the institute's total funding, or approximately $50 million annually.

The Early Genetics Program

Although NIGMS, from its beginning, had a clearly interrelated set of activities supporting genetics research, it did not achieve the formal status of a program until 1972. However, we will refer to all these activities as the Genetics Program. The early history of genetic research lacked social contentiousness. To the public, genetics was a rather obscure, although possibly important, field of research remotely related to the alleviation of disease and suffering.

The early real interest in genetics research was limited to the scientific community which conducted the research and the political community which funded it. The relationship between the institute and its constituencies was relatively simple. The sole commitment within the scientific community was to conducting basic genetic research; it clashed with neither society nor government. Funding came from the political community. Although its inability to understand the details of genetic research might have hindered appropriations, the high status of genetic researchers and recent dramatic discoveries dominated the thinking of key political figures and the necessary funds were provided. Men like Senator Hill and Congressman Fogarty were central to the phenomenal growth in financial support of NIH between the end of World War II and the early 1970s.

Although, to a large extent, the program dominated the political community, the program itself, in turn, was dominated by the ideas of the scientific community. First, there was the strong bias toward "pure" scientific research which favored "basic" over "applied" projects; a long-term orientation over a short-term one; the individual researcher-initiated project over larger-scale contract research.[2] There were strong beliefs that the scientific world should be self-stimulating in terms of choosing research topics and

---

[2]D. Stetten, J. Shannon, P. Handler, L. Thomas and N. Anderson. *On the Stewardship of Basic Science and the National Institute of General Medical Science,* remarks on the tenth anniversary of NIGMS, 1973.

priorities; that administrative organizations should refrain from intervening in the business of science and make no attempt to direct it; that project funding should be based on scientific merit and not the needs or demands of groups in the political environment. Autonomy was highly valued by the scientific community: scientific affairs were best left to the scientists and the informal workings of the scientific community.

Since the program's major constituencies were essentially in synchronization, there was no need to develop a differentiated organizational logic, and there was virtually no distinction between it and the program's political logic. Both were governed by the values and ideas of basic research scientists. The internal administrative structure was organized into sub-units by mechanism of support: research grants, research fellowships, and research training. This was designed to support the growth of the scientific community itself. Research training produced people capable of doing research in genetics; research fellowships supported them in their post-doctoral work; and research grants supported the continuing activities of research scientists.

The institute had a problem, however, in choosing exactly what research projects to support. Officially, it could support anything that was not a primary responsibility or interest of another institute. In fact, it chose to follow the advice of study sections as its guide. Study sections are a device for evaluating grant applications. Applications arrive at NIH and are sorted by their scientific discipline and allocated to study sections for genetics, endocrinology, immunobiology, etc. These study sections, staffed by external expert academics, evaluate and rank individual grant applications using scientific criteria. A ranking is sent to the institute, which generally funds projects according to the assigned priority and the availability of funds.

In its early years, NIGMS dispersed funds according to the interest, priorities, and evaluation of the scientific community. Study sections, staffed by scientists, viewed a range of research projects by other scientists. The values of autonomy and self-direction were operationalized through this administrative mechanism. Funds were allocated primarily as grants to individual researchers for their independent research projects, rather than as contracts for a deliverable research product. The result was a close fit between this type of research support mechanism, the limited knowledge of the physical phenomena, and the use of stage 1 and 2 technical logic. Both the scientific and political communities were satisfied. This early period developed a cadre of genetic researchers and a blossoming of knowledge. Table 4 summarizes our analysis of this early period of the Genetics program.

External guidance, intervention, or use of non-scientific criteria, were irrelevant as long as the driving forces in the outside world were moving in the same direction as that desired by the program. This was the case until the cracking of the genetic code in the middle 1960s.

TABLE 4
EARLY GENETICS PROGRAM

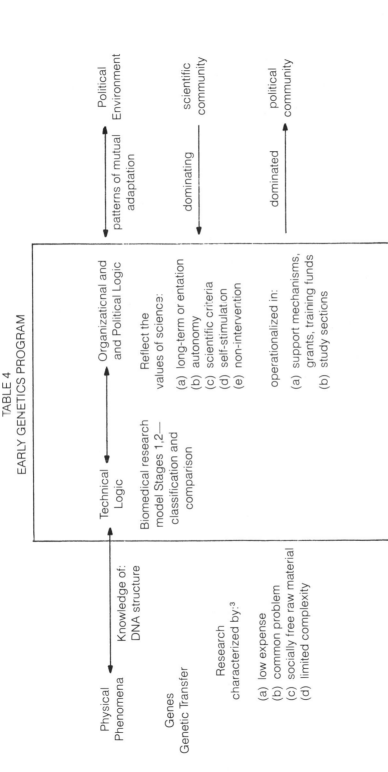

[3] See Chapter 3 for the discussion of these research characteristics.

The Changing Context of the Genetics Program

The relatively simple environment just described changed with the discovery of the genetic code. For a long time genetic research activities had been categorized into two types. One "stream" was genetic chemistry, a subset of molecular biology. The other, human genetics, focused on specific genetic diseases and more applied issues. The major effort prior to 1966 went into the former, more fundamental "stream," reflecting Watson and Crick's opening up of a whole range of research questions at the cellular level of analysis. The cracking of the genetic code provided important knowledge for attacking the complex problems of human genetics, and after 1966, research emphasis started to swing in this direction.

The drama of this discovery was a major factor in awakening the public to the importance and enormous potential (positive and negative) of genetic research. Claims that genetic-based interventions potentially could alleviate suffering from genetic diseases fueled great interest and the formation of numerous constituency groups advocating research directed towards particular diseases. The number of letters from political and social groups expressing an interest in genetic research increased dramatically in the years after the cracking of the code. There also was an increase in communications expressing concern about the dangers of genetic experimentation creating new, unknown, potentially dangerous organisms, and about ethical issues of genetic research. The social/political environment became more complex with the emergence of special-interest groups and multiple value sets.

The technical success of genetics research led to difficulties within the scientific community, as well. Many new research possibilities were opened up and a proliferation of research proposals followed. This trend was attributed to several developments:[4]

1) an improvement in technology, particularly in tissue culture genetics;

2) an increased desire and ability to extend and modify conceptual frameworks developed on lower model organisms to human genetics;

3) a realization that such studies were of direct relevance to current medical problems, and

4) increased inter-disciplinary interactions, particularly among biologists, bio-chemists, geneticists, and research clinicians.

For the first time, the Genetics Program felt a need to identify areas of concentration to ensure a continuity of research and to maintain control over the program's future direction. A new mode of adaptation was required to resolve these issues and move the technical logic from the early classification and comparison stages to the mechanism stage and a focus on human genetics.

---

[4]NIGMS Annual Report, 1972, p. 61.

A Second Mode of Adaptation

The 1968 NIGMS annual report reflected some key changes in language and concepts used to direct the efforts of the institute:

> In summary, NIGMS adopts the proposition that a problem-solving, mission-directed orientation in program administration is more productive and rational than the former passive grant by grant disciplinary approach. . . .

What was occurring in NIGMS as a whole was the emergence of a new dominance within the organizational logic of the program: a transition from the "basic" or "fundamental" science orientation to a "basic-targeted" logic. A major feature of this transition was the development of new language within the institute consisting of such terms as "relevance," "targeting," "programming," and "problem-solving." This language extended the existing logic so that it became selective while remaining basic and fundamental. It was fundamental research because it still drew upon certain disciplines, language and theoretical structures; but instead of allowing scientists to roam free across the original theoretical territory, activity was starting to focus in selected key areas.

The importance of bringing together the ideas of targeting and basic research should not be underemphasized. In the scientific community and within NIH, there often is dichotomous thinking about basic research and applied or targeted research.[5] A House of Representatives staff report[6] on the distinction between contracts and grants as means of research support exemplified this point:

> Early in the investigative staff study it became clearly evident and was repeatedly confirmed during interviews with research scientists that there exists in the biomedical community a marked difference in philosophy concerning research. It plays a role in the expression of opinions, prejudices, and certain apprehensions concerning areas germane to this study. It relates to what appears to be a continuing philosophical tug of war between the "PhD and MD types" involved in research, between protagonists of "grants" versus those of "contracts."

> Illustrative of this issue is an anecdote by an NIH official. When a "grants" advocate boasted to a "contract" man that he did not know of a single Nobel prize ever having been awarded for work done under a research contract, the

---

[5]This dichotomous, adversary thinking is not unique to NIH, nor is it an especial criticism of NIH. This difference was observed by Merton (1968) in a major work on the sociology of science in the Western world in which he discussed in detail the conflict between two extreme logics—science and a totalitarian form of goverment.

[6]Hearings before the appropriations Sub-Committee of the House Committee on Public Health and Environment for Fiscal Year 1974.

"contract" supporter retorted that he was unaware of anything like an artificial kidney having been developed under a research grant.

The same report quoted the perceptions of an NIH official about the culture within NIH:

> He pointed out that most scientists on grant application reviews consider much contract research to be routine, pedestrian, a type of "scientific prostitution," invading upon what they might consider the essence of bio-medical research, namely the unrestricted and undirected pursuit of basic knowledge.

These distinct orientations, basic versus applied research, can generate disruptive win/lose conflict or a useful tension as we think happened in the Genetics Program, through developing a synthesized basic-targeted logic and language. Administrators had made a fundamental accommodation in organizational logic and strategy while continuing to grow and to achieve scientific results. The fundamental research commitment and credentials of the institute were not called into doubt by the scientific community. Its origins and achievements presumably would have convinced any skeptic of its scientific "purity." We think conflict was avoided partly because the language of targeting and programming was directed toward the new state of knowledge of the physical phenomena, while the language of basic science was still directed toward ultimate purpose.

All research applications, however, still came through the study section review procedure where traditional scientific disciplines predominated. Proposals concerned with human genetics tended to be downgraded in this process. This was creating a misfit between the existing research support mechanisms and the emerging interest in developing further understanding of human genetic disease which the state of knowledge now permitted.

To cope with these developing misfits, NIGMS became more proactive in its relationship with the scientific community. It developed new task support mechanisms like Centers for Human Genetics, congruent with the new task realities. The director of NIGMS, Dr. DeWitt Stetten, commented on this change:[7]

> Although the individual project grant remains the mainstay of NIGMS's support in the fundamental sciences including genetics, larger projects and centers will be required to stimulate and advance research in human genetics. This is particularly true at the clinical level which requires the cooperation and integration of individuals in a variety of disciplines and skills. . . . Cooperative ventures will be needed in order to make progress in three of the four NIGMS initiatives in human genetic disease; the mapping of the human chromosome, early identification of human genetic disease, and greater understanding of the genetic control of human cells.

These centers were conceived as facilities to translate advances in basic genetics into clinical application. They would involve a community of basic and clinical scientists doing research and providing community service.

---

[7] NIGMS Annual Report, 1971, p. 5.

After defining the center concept, NIGMS staff actively encouraged submission of proposals for large center grants. There was a conscious effort to lead the scientific community in the direction of two objectives: first, a rapid extension of concepts and techniques derived from the molecular genetics of simple organisms to the level of human genetic disease; second, bringing human genetic disease to the bench of the bio-chemist and molecular biologist for the type of sophisticated scientific analysis previously limited to lower organisms. The latter objective was in turn supported by development of a special bank or "library" of human cell tissue cultures carefully indexed and made available to researchers under controlled conditions. This intervention was needed because the free market of scientific endeavor could not have produced such a facility at a reasonable cost. It was too expensive an undertaking for any individual scientist, and the level of cooperation and coordination required the help of some outside agency. In addition to new task support mechanisms, NIGMS continued to evolve its programmatic structure. For the first time, in 1972, a true genetics program emerged.

A number of factors seem to have contributed to the actual reorganization. Although it would be misleading to imply direct causality, breakthroughs in genetics research were a factor in the organizational changes. A second element probably was the desire within NIGMS to develop into more than a residual institute. The old organization structured by funding instrument was losing support at NIH. A special committee, the Cooper Committee, was arguing for a move toward the concept of program organization. The new director of NIGMS came from an administrative position in a university and the new organization reflected the disciplinary structure found in universities. Finally, although not a prime motivational factor, President Nixon impounded all research training money in 1971. This action eventually was overturned by the courts, but in 1971 it appeared that the training branches of the institute would have to be phased out, necessitating a reorganization and reallocation of responsibilities. In any event, Dr. Fred Bergman, who became the Genetics Program manager in 1972, believed that reorganization was a deliberate attempt to increase control over the future direction of the program and its allocation of funds.

The picture that emerges in 1972 is one of simultaneous changes in the state of knowledge, in society's involvement in genetic research, and in the internal administration of NIGMS and genetics research activity. The result, whether or not specifically articulated and planned, was an administrative structure adapting to the new task and political realities.

Managing the Genetics Program after 1972

Dr. Bergman, in his 1974 annual report, defined the ultimate goal of the Genetics Program as the prevention, therapy, or cure of human genetic disease. The means to this goal were the support of "excellent research and training in the broad, overall area of genetics." In that same report, he commented that those non-controversial statements about goals and means did not address the complex issues inherent in the program or face squarely the fact that the program could not exist in splendid scientific isolation. His program served a social purpose. Brief examples of some of his actions will show how he managed the program and positioned it in relation to both science and society.

The Program's Role in Science

Dr. Bergman described his function as program planning or the process of devising and implementing a strategy to eradicate genetic disease. Ninety-five percent of program time was selecting and monitoring for the excellence of funded research. The remainder, of increasing importance, was acting as a catalyst for new program initiatives.

To this end, Dr. Bergman developed and maintained a mix of research activities (basic and applied) or "portfolio" of efforts linking basic scientists and clinicians. The program was heavily "model system," long-term discovery oriented (70 percent), with some effort (30 percent) directed to short-term oriented exploratory activity. He strongly believed in "funding the best, the highest quality research" but, even with his predominant orientation toward investigator-initiated projects, Dr. Bergman had taken steps to become more active in the stimulation of proposals:

> If I have a radically new idea, I'll get advice from outside consultants and write a white paper for circulation. I'll become an advocate for the idea and stimulate submission of proposals.

He also would use his limited discretion "sporadically and cautiously" to fund proposals that weren't at the top of the scientific study section's priority list if he felt they met an important, unaddressed need. He believed reviewers generally ranked "basic" research proposals better than "applied" proposals and thus ran counter to his concern for increasing the quality of clinical, human-oriented genetic research. Dr. Bergman had staff members analyze the 136 grant proposals reviewed for funding in 1974 and categorize them by their orientation. The results generally supported his contention:

> The analysis supported the idea I had that technical reviewers of grant applications referred to the NIGMS Genetic Program tend to be more enthusiastic about research in basic genetics and molecular biology than about the relatively fewer—but growing—number of applications which are more directly applicable to problems in human genetics. The former are often described as "elegant" and 'promising'; the latter tend to be described as "competent," "fraught with artifacts," and "difficult."

It was not always the reviewers, however, who could be blamed for the low priorities. Some of the more "applied" proposals just were not satisfactory. Once having identified this problem, Dr. Bergman took action, such as sponsoring a workshop to improve research on the genetic counseling process. A two-day workshop addressed issues like the goals of the studies, potential study biases, use of control groups, and collaborative studies. Dr. Bergman explained:

> During the last two to three years, a number of research grant applications have been submitted to the NIH proposing to evaluate the effects or the effectiveness of current counseling practices, clearly an important and timely research topic. However, the proposed projects were frequently poor in quality and often lacked sufficient input from medical sociologists acquainted with the complexities of attitudinal research. Consequently, such applications usually received low priority and were not funded.
>
> In view of a clear need for improvement, program staff discussed the matter with the Genetics Study Section and later, in March 1974, with the National Advisory General Medical Sciences Council. We subsequently planned and sponsored a workshop.

Another area of increased program activity was balancing the ratio of large grants to small grants. In 1971, NIGMS had started encouraging large "program project grants." Often referred to as "umbrella" grants, they aggregated many projects in a single institution. Large grants of this type were supposed to be synergistic: interdisciplinary interaction within the funded organization making the whole greater than the sum of the projects. However, the high degree of collaboration often did not exist. Also since they usually extended over a full five years, they could more easily avoid review. Dr. Bergman commented:

> We used to be lumpers and not splitters. Often mediocrity crept in. They often would issue purposely long and confusing reports to avoid adequate reviews. You have less control over them and they tend to take over.

The program reviewed its procedures for evaluating large grants and looked more closely at the interactions. Program officials began pushing questions like:

- Does the program project make sense? Should there be one large grant or ten small ones?
- Does it have scientific merit?
- What is the pattern of collaboration?
- Are the grantees working hard and producing some interesting output?

Dr. Bergman would review grant proposals for program projects and recommend smaller independent projects if he thought that action appropriate. He also would make site visits when problems emerged or when the five year funding was up for renewal. On occasion, the program phased out support of large grants when there was insufficient coordination or leadership.

Finally, we would mention the mixing of day-to-day operating problems with the broader policy perspective necessary to manage a program actively. Some staff members believed that Dr. Bergman's responsibility was policy and theirs the day-to-day operating activity related to research grants. Dr. Bergman disagreed and involved his staff in policy matters as well, often by having them sit as members of various policy committees like the mutant cell bank committee or the recombinant DNA committee. He, personally, also kept close to operations, as well as concerning himself with policy matters.

The time division between operating and policy activities approximated a 60/40 split for himself and his staff, with the larger amount of time spent on operations. One could argue whether a 60/40 split in administration and a 70/30 split in portfolio mix were appropriate numbers, but it cannot be argued that this manager failed to establish some criteria to guide action toward a defined and desirable end—the prevention, therapy, or cure of human genetic disease. Moreover, genetics was a relatively small program, making this mixing of policy and operations more feasible. As the size of the organization increases, a more traditional separation of policy and operating responsibilities by hierarchical level such as Bell's appears.

The Program's Role at the Boundary of Science and Society

The social and political environment of the Genetics Program became increasingly volatile as genetics research became a visible social issue. Social constituency groups mushroomed, each expressing its concern about a particular genetic syndrome or disease. There obviously was great potential for direct confrontation between the demands of these groups and the limited resources and research orientation of the Genetics Program. The commitment of the program to the alleviation of suffering from genetic disease required that new knowledge, particularly about intervention and counseling, be transferred to the health care delivery system.

The program administrator found himself more involved with external groups for two reasons: social consciousness and program integrity. He commented:

> Our knowledge is terribly important. We are a knowledge-generating business. When you see that existing knowledge isn't being used, you are in trouble. I am concerned about that. You need to remember the ultimate goal of the Genetics Program is the prevention, therapy, and cure of human genetic disease.

Although committed to alleviating suffering from genetic causes, he also cared about maintaining program integrity. The Cancer, and Heart and Lung institutes had become involved in large demonstration and service programs, using money that otherwise could have been earmarked for research. He was anxious to avoid a similar situation:

It is in our interests to encourage others to disseminate the information. I believe it is right to get these things going. I see myself as a catalyst. This (dissemination of information and health care delivery or service) is our business, but not our appropriate role. Unless we can stimulate other agencies, outside interests will pressure us to do something that we don't have the resources to accomplish.

There was a clear example of the program administrator distinguishing between being a doer and a catalyst, thereby defining and managing the boundaries of the Genetics Program around the issue of genetic counseling. Dr. Bergman explained the situation:

The primary role of the program is biomedical research. Although research into the delivery of certain health care services, such as counseling, is appropriate, it focuses on the extreme end of the spectrum of activities that I consider part of our mission. It should be of greater concern to the federal agencies involved in health care delivery, such as Health Services Agency. Greater coordination with other agencies will help us to define our roles more clearly.

He arranged meetings between himself and representatives of organizations concerned with health-care delivery.[8] He was explicit that this process was intended to define a boundary. The desired relationship with health-service delivery was one where actual service was outside his organizational boundary, but where he was linked with it and provided inputs to it. He illustrated his concept as shown in Figure 26.

The program administrator recognized that the public's interest in genetics would continue to increase and he was developing educational materials for dissemination:

We consider it one of our important functions to provide information on genetics in general to the public. We prepared and published a 30-page brochure, "What are the Facts About Genetic Diseases?" The booklet has been well received and is being used in a number of high school and college courses, and by volunteer organizations concerned with genetic disease. We are currently developing another publication which will focus on research progress in medical genetics.

The program also had to contend with a change in relationship to the political community. Total NIH budgets have, if account is taken of inflation, declined since the early 1970s. The deaths of Congressman Fogerty and Senator Hill and an increasing skepticism about the output and role of scientific research combined to put increasing pressure on the NIH budget. Reorientation was required by the scientific community, as well. Whereas the objectives of the scientific community in pursuing basic research once

---

[8] Such representation included the Health Services Administration, Health Research Agency, psychologists and sociologists, the National Genetics Foundation, the Institute for Society, Ethics, and Life, and the Cystic Fibrosis Foundation.

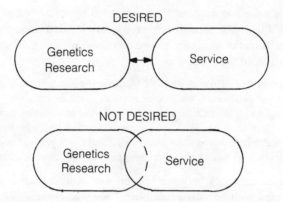

FIGURE 26
GENETICS PROGRAM BOUNDARY PLACEMENT

were identical with the Genetics Program, now there was room for considerable divergence. The need to pursue human genetics with its less clearly defined and disciplinary-oriented problems required that the genetics program stimulate changes within the scientific community and handle the delicate problem of the slow reaction of mechanisms like the study groups.

A considerably changed set of social/political realities existed in the middle 1970s than in the early 1960s. However, changes in the program's organizational logic to incorporate ideas like targeting and intervention led to a more symbiotic relationship with all the elements of the program's environment. The new set of relationships is diagrammed in Table 5.

The very success of the first mode of adaptation contributed to the emergent problems and misfits which had to be resolved by creating a second mode of adaptation. Both modes of adaptation were characterized by clearly distinct organizational, political and technical logics. These two adaptive modes are summarized and compared in Table 6 below.

Organizational adaptation is a complex and lengthy process brought on by the necessity to respond to changes in the technical and political environments. The Genetics Program provided a useful opportunity to trace this process. It offered two clear and distinct modes of adaptation, linked by a process of change and organizational elaboration. We see the need for a framework and language to explain this process, to increase our understanding of it, and to facilitate the development of normative ideas about successful adaptation. Our conceptual framework permitted us to order the events and changing relationships in the history of genetics research and we believe it will lend itself to critical analysis of other R & D programs.

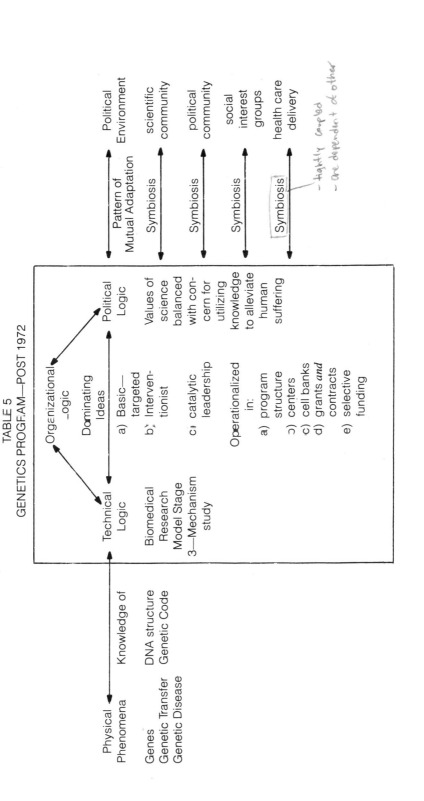

TABLE 5
GENETICS PROGRAM—POST 1972

TABLE 6
COMPARISON OF TWO ADAPTIVE MODES

Conceptual Element	Mode 1	Mode 2
1. Knowledge of Physical Phenomena	one problem	many puzzles
	low cost	increased cost
	common focus	divergent focii
	socially free material	encumbered raw material
	limited complexity	increased complexity
2. Technical Logic	Stages 1, 2	Stage 3
3. Political Environment	uni-dimensional	multi-dimensional
4. Political Logic	Values of science	Values of science balanced with responsiveness to socially articulated issues
	basic research	
	self-stimulation	
	autonomy	
	scientific criteria	
	passive	
5. Pattern of Mutual Adaptation	dominating and dominated	symbiotic
6. Organizational Logic	same as No. 4	basic-targeted
	individual project, grants, funding on study section recommendation	interventionist
		catalytic
		cooperative projects, centers
		program structure, cell bank
		grants *and* contracts, selective funding

Chapter 8

Managing Large R & D Programs: Summary and Conclusions

Our findings from the study of selected research and development programs at NIH and AT&T have been presented in the form of a model: an integrated set of concepts that replicate and clarify a very complex process. We have used the model to describe and order the events of the programs and to uncover patterns of behavior and organizational arrangements that contributed to their success or failure.

The model developed as we asked ourselves how a senior R & D manager could order the facts of the programs to serve as a useful guide to management action in the future. So our model, as a first priority, has been created to guide R & D management practice, even though we believe it may have a wider utility. A model so designed cannot be rigorously "proven" and we make no such claim for it. Its only "proof" will be from future evidence of its usefulness in helping R & D managers order the realities of other programs. Many programs are more modestly scaled than the ones we studied. Does it help in understanding these smaller programs? Does it prove useful in technical areas quite different from bio-medical and telecommunications research and development? Can it provide a basis for additional and more refined learning about the R & D process?

For now we can only say that, for us, the model does what it was designed to—help in understanding and accounting for the facts, as we know them, about our sample of programs covering a considerable range of R & D topics in two very different organizations. It does this with an economy of concepts in the interest of the simplicity that will make such a model useful. The model, we believe, can become a working tool for the senior R & D manager: an improved theory of what it takes to achieve desired outcomes in the face of the inevitable political and technical uncertainties of the R & D world.

The R & D Management Model

Early in the project we became aware of the political environment's potential impact on research activities at NIH. Beyond the financial support necessary to maintain the organization, there were issues of how best to manage research and even which research to undertake. Congress, the Executive Branch, and special interest groups championed programs that proved expensive and fell short of the espoused goals despite worthy causes, good intentions, and serious efforts. In talking with people involved with some of these externally initiated programs, or in reading historical documents, one got the feeling that an attitude often prevailed that the major obstacle to curing a disease was the lack of money; that nature would yield its secrets in the face of a monetary assault. In some cases, as proponents of this line of reasoning can demonstrate, it has worked. If it has been done once, why not again? If we can put a man on the moon, surely we can cure the common cold, heart disease, cancer, or whatever else ails us. But as we have seen, this approach does not always work. We are faced with an array of outcomes—some positive, some not—and we must wonder why.

One reason, probably valid at times, is that individuals or groups in the political environment do not understand the intricacies of the subject matter, and exceed their areas of expertise and judgment. But political intrusion cannot be blamed for all problems. We have seen that researchers, themselves, can be trapped by their own optimism and desire for tangible results. So, although elements of the political environment may have pushed researchers toward more applied and developmental work and into activity beyond the existing limits of knowledge, intrusion did not explain all program problems.

Even without political intrusion, the management of research and development is a complicated task. To discern why some programs were less successful than others, we had to consider some technical aspects of the research and the research process. To allow us to classify and compare our diverse set of programs, we developed the concept of the technical logic of R & D. Once we understood where each program was positioned on that continuum, we looked for patterns of success and failure. Using the concept of the technical logic, we found instances of programs either choosing or being "forced" to start at inappropriate stages. In distinguishing between the sequential and empirical research strategies, we recognized that, in some situations, it may be possible to find cures or develop systems without the benefit of a complete discovery phase. This was not the case for the programs in our sample, however. Beginning at the wrong stage and/or failing to switch to a more appropriate stage resulted in expensive programs that failed to achieve their goals.

The senior R & D executive also must interact with a dynamic political environment and manage this relationship. If the fruit of technological de-

Managing Large R & D Programs: Summary and Conclusions

velopment impacts society negatively, special interest groups and elected representatives may attempt to influence directly the research process more actively. What was a complicated job previously becomes even more complicated as the political environment increasingly makes itself felt in more organizations—public and private. R & D organizations must face this growing interdependency between their political and technical environments and respond by first recognizing its possible inevitability and, second, by actively managing both environments. They must continually adapt, or coalign, their organizations to their multiple contexts, so that neither a political nor a technical logic dominates to the detriment of the organization or society. Executives in the successful programs we studied managed this adaptive process well.

How does a senior research executive adapt his or her organization? What factors deserve special emphasis and consideration? We have identified four such areas of executive action from our case studies: 1) positioning the logical, as well as physical, *boundaries* of a program; 2) *switching*, as needed, political and technical logics and the resources to fit prevailing conditions; 3) *synthesizing* conflict between the political and technical logics, and 4) establishing relevant *criteria* for performance review. We see the performance of these functions as one way to think of the general management role in an R & D organization. Figure 27 summarizes our model, including these four dimensions.

FIGURE 27
R & D MANAGEMENT MODEL

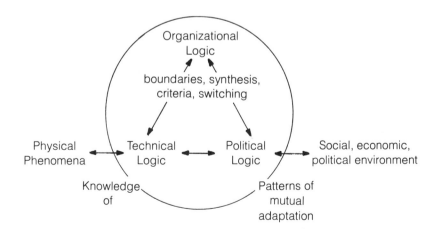

Managing Boundaries

Both physical and logical boundaries exist within an organization, as we saw in AT&T, and between the organization and its environment. Distinguishing physical boundaries is not particularly difficult, for one can see buildings and other assets such as equipment and supplies, or follow a flow of inputs through a transformation process and observe the output and exchange. But the substance comprising the logical organization, a conceptual demarcation of orientations, interests, and responsibilities, is far less tangible; accurately defining logical boundaries is difficult.

Research executives must attend to the process of defining and managing boundaries to ensure the continued integrity of their programs. It is a process of deciding what is done well within a particular area or program and what is better done elsewhere. Over-responsiveness to environmental demands for service activities, for example, can blur a focus on discovery and thereby inhibit development of cures or treatments. Passivity, on the other hand, can result in an overruling of the research manager's boundary placement, or in a disrupting reduction of resources. Where resources are limited, boundary management aids direction of program activities, like emphasizing discovery, rather than application, or vice versa. It is a process of focusing attention and skills where they are best used, but not unilaterally ignoring other legitimate concerns. Trying to solve all problems simultaneously can be impractical and very expensive.

We have seen how one organization, Bell Labs, manages its internal boundaries in order to move knowledge from discovery through to development. The conscious attention to boundary management was most obvious in the use of branch labs as a device to spread out and modulate the tension between the logics of research and high-volume manufacturing so as to encourage a learning relationship between them.

Social and political issues often arise when the technology begins to affect the public. The concerns of groups in the environment undoubtedly always are present, but possibly not attended to until the new technology is ready for use. A danger exists if technologists ignore the impending reactions of the political environment or wait too long to deal with them. Resolution of social uncertainties as well as technical ones is essential to a program's success. Placing a boundary too closely around the technical task potentially ignores the political reality in which a research program exists. Encompassing too many environmental demands, on the other hand, can diffuse the technical effort.

In the Sickle Cell Anemia Program, boundary establishment never was much of an issue after initial NIH attempts to ignore or avoid the program failed and a highly service-oriented program was thrust upon it. Only the deputy director of the Heart and Lung Institute was willing to negotiate a boundary closer to the societal demands, which clearly were the critical contingency in this program's early stages. The boundary ultimately was established by the secretary of the Department of Health and Welfare when he appointed a program advisory committee reporting directly to him. The committee's objectives were divided between research and community service. Unable to agree on priorities, however, the committee divided the available funds equally between research and service, thus diffusing the level of effort of each. NIH had the responsibility for coordinating a substantial service effort which few people at NIH seemed to support, even though the service responsibility was placed well within NIH's boundary.

While the Sickle Cell Program did not concern itself with setting a boundary, there is no question that the manager of the Artificial Heart Program had a clearly defined one in mind. The domain he staked out was drawn tightly around the technical task of developing a fully operating replacement heart, even though the scientific knowledge and development expertise were not available to support such an undertaking. The critical contingency facing the program was not the public, but rather the director of NIH, who voiced the concern of the scientific community that more work at the discovery and exploratory development stages was necessary, and an administrative concern that a large scale development program would drain funds from other vital projects.

Essentially, administrators of the Sickle Cell Program did what they were told, and those of the Artificial Heart Program did what they chose. The essential positive tension at the logical junctures of these programs and their environment was ignored or never materialized. On the other hand, we saw the SIDS Program administrators respond to the tension positively in their interactions with SIDS parents and Congress.

Managing the program-environment interface appeared as an issue in more NIH programs than in Bell programs, probably because of NIH's polycentric structure and openness to the environment. We did see in the high capacity mobile telephone system case the institutionalized way that the Bell System managed this interface when necessary, however.

As key environmental uncertainties shifted, Bell recognized the change and responded by placing specialist groups to manage their important interfaces. The first shift occurred when the FCC indicated a willingness to expand spectrum allocations. At this point, the critical contingency was technological, and control of the situation passed from AT&T marketing and engineering groups to a technology focused unit with the creation of a mobile communication department within Bell Labs. The AT&T units had performed their function of monitoring the environment and mapping it into the Bell System prior to Bell Labs' taking over.

With the advent of controversy surrounding HCMTS, the interaction patterns of the Bell System again shifted to meet the contingency. The key uncertainty no longer was technological, but rather political, and AT&T again moved in specialist units. Bell Labs no longer interacted openly and directly with the FCC, but had to go through mediators. The Regulatory and Legal divisions obviously were better qualified to deal with the new environmental conditions and interactions with the FCC and courts which ultimately would define the new ecological balance.

Adaptive organizations manage these internal and external boundary relationships quite explicitly, ensuring, as best they can, continued survival in complex, shifting political and technical environments. The process of boundary management is largely a means of adjusting the tension that is brought to bear on an organizational unit. Tension can be either too high or too low to induce the learning that is needed in high performing R & D programs. This relationship is schematically shown in Figure 28, along with references to two examples of when shifts in tension induced renewed learning. In the cancer chemotherapy case, it took a significant cut in funding to move the program out of its complacency and shift its technical logic. In the case of artificial heart, the great tension caused by the breech between the program manager and the NIH director had to be eased by changing personnel before the program began to learn its way back from the frozen logic of its "master plan."

One organizational theorist (Thompson, 1967) stated that boundaries tend to extend in the direction of critical contingencies. We would add that one first must identify these contingencies. Some of our case studies indicate that the successful programs do, in fact, make this assessment and act as the theory indicates. But it is a difficult exercise, not automatic, and the less successful programs experienced difficulty with the process.

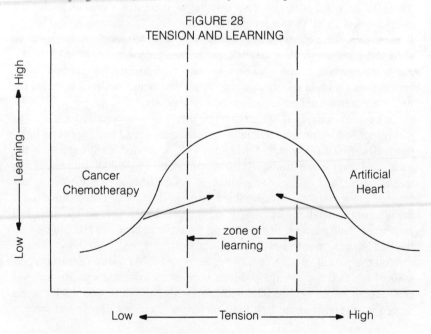

FIGURE 28
TENSION AND LEARNING

Creating a Synthesis

It is difficult in R & D organizations continually to balance the administrative dictates of research, development, and implementation. Single logic domination can contribute to the destruction of a program's integrity by preventing the requisite fits with the political and/or technical environment. For example, attempting to introduce the logic of researchers or design engineers, who create new systems and strive to utilize the newest technological findings, into a mass production operation conceivably could overwhelm this process which, to be successful, must find ways of reducing variety and increasing standardization. This mix would tend to push the mass production unit in the direction of a unit production job shop and thus undermine its original logic. The unit no longer would remain efficient in its technical task. Similarly, it may not be sufficient to solve technical problems efficiently and then worry about implementation or marketing problems. It should be a simultaneous process and not strictly sequential. The concerns of the organizational units near the implementation end of the innovation process need early representation.

It takes a special quality, or role orientation, for senior R & D managers to recognize the appropriateness of such contrasting rationales in order to reconcile them. Beyond understanding, however, differing viewpoints must be valued so as not to simply capitulate to the pressures exerted by one against the other. A manager must create a synthesis, or new logic, which is operationally consistent with the "truths" of each opposing logic. This is simultaneously an understanding and valuing of opposing positions and an orientation to action to resolve debilitating differences. We identified a type of leadership that seemed to contribute to the establishment and maintenance of program integrity—dual advocacy.

Dual Advocacy

Dual advocacy can be seen as a dialectical process; a process of creative readjustment of interacting systems and the development of a new equilibrium. We prefer to think of it as a mode of resolution in which: ". . . the synthesis not only subsumes and retains elements of the thesis and antithesis, but creates new patterns, structures, and attitudes, as well. The synthesis results from the creative opposition of dynamic forces and is contingent on some perceptual, conceptual, or empirical reorganization by both the organization and its members" (Lourenco and Glidewell, 1975, 491).

The R & D administrator's role as a dual advocate can be summed up as being an advocate for the values of the technical logic in relationships with the environment; and an advocate for the values of the political environment in interactions with the researchers. It is the process of ensuring organizational integrity—searching for opportunities in the environment and holes and failures in the technical logic. Only by stepping into these voids can the R & D administrator or policy executive maintain the creative balance between science and management. The environment will continue its press on R & D organizations and the dual advocate must maintain the value of good science, while being responsive to social needs.

In his or her concern for organizational integrity, the administrator becomes concerned with the whole innovation process, sees the interrelationships which comprise the gestalt, and becomes committed to his or her role in the process. The dual advocate realizes that no important problem or undertaking is so simple that the answer lies within one logic.

The dual advocate serves to promote a synthesis—the creation of a new situation or substance out of two inputs having different identities. It is not an easy role. It would seem to demand a number of attributes:
- a clear perception of the identity of the organization, the aims of which one is trying to further and the integrity of which is to be protected;
- a clear perception of the environment's complexity;
- an ability to identify and relate to centers of power in the environment;
- a boundary-spanning capacity to see and respect one's own logic and others' without being totally captured by either.

When two different, but relevant, value systems clash, progress is likely if polarization is avoided. If the two divergent systems can be synthesized, then new energy is likely to result. The resolution should be a genuine synthesis and not just a compromise or smoothing of the situation. Confrontation may be a useful way of avoiding debilitating compromise, but it, too, has limits. The symbols by which people live are very close to their identities and significant changes can be very painful. Reassurance that their best interests will remain unprejudiced by the new logic is important, and strategies should be constructed to make this principle explicit.

At NIH, this style seemed to describe best the leadership of the SIDS and the genetics programs. Without repeating all the details, we can say that in these cases the leadership was able to learn from and, in turn, to influence both their political and technical environments. For example, in SIDS it was important to know that a discovery effort should be the primary thrust, but at the same time to learn from parents and Congress that people were very upset and wanted everything possible done. Using this learning, the leadership then could influence and advocate in both directions. They clarified to parents the real limits of knowledge and therapy and to Congress the limits of useful funding. At the same time, they communicated to the scientific community the social importance of the problem and the possibility of fresh scientific approaches. It was through this complex process that the program enhanced its integrity during this difficult period when it easily could have been discredited.

That the program administrator of the Genetics Program was acting as a dual advocate needs little comment. He espoused the values of science and social responsibility and took action on his beliefs. He believed in "funding the best research"; yet, when he saw a need that was not being addressed, he had the flexibility to fund relevant, lower-rated projects and provide for workshops to upgrade research in emerging areas of important social impact. He also functioned as a catalyst in stimulating service agencies to disseminate and use information produced by the program.

Some organizational theorists, Thompson (1967), Kotter and Lawrence (1974), have suggested that a major role for administrators is co-aligning the organization with its external constraints and environments. We think the concept of dual advocacy goes beyond the notion of co-alignment and requires further examination, particularly the initiating, catalytic aspects and the transfer of values across boundaries.

The Genetics Program, like SIDS, was relatively small, not highly differentiated, and did not require complex administrative mechanisms. An individual with the dual advocacy orientation was sufficient to manage these programs effectively. At Bell, we saw examples of how program integrity was maintained in programs that were too large to rely completely on the personal leadership skills of one or two people without the assistance of administrative devices. In these complex programs, multiple groups from scientists to customers were tied together by an array of administrative mechanisms chosen for each particular situation. The set of task-force memberships, computer linkages, organizational bonds, cross-functional and cross-company committees, temporary assignments, development organizations, and systems engineering served to foster, support, and legitimate the integrative concern that is the heart of dual advocacy. These mechanisms, of course, could not have operated without skilled leadership, but individual leadership, by itself, could not have carried the necessary integrative effort required.

A program manager's orientation is undoubtedly important but it needs support in the form of institutionalized administrative mechanisms to ensure the maintenance of adaptive conditions. These administrative devices do not ensure that the right decision or adaptive actions will be taken for, in the final analysis, that is an act of judgment. Managers must resolve the contrasts and contradictions thrown up by divergent viewpoints. The administrative devices function primarily to ensure that all relevant premises will be presented as input to the decision process, thus minimizing the possibility of misfits either with the environment or in the technical logic.

The cases have shown clearly the importance of having well-prepared and administratively well-supported managers to achieve the richness of integration required for successful R & D programs.

Establishing Criteria

Research programs must establish and maintain boundaries appropriate to their role in diverse environments and often synthesize competing logics that threaten their integrity. To achieve this synthesis, however, the organization must possess some criteria for setting priorities, evaluating work in progress, assessing the quality of relationships with other social groups, and serving as a guide to action. At one level, the establishment and use of criteria are heavily scientific and technical and concerned primarily with operational issues of developing knowledge, cures, or hardware systems.

At Bell, we observed that objectives and criteria were congruent with the stage of the technical logic at which a program was positioned. At each stage the objectives, criteria, and evaluating groups shifted as the knowledge or system was "handed-off" to another organizational unit. AT&T delegated authority to Bell Labs to choose discovery projects and to monitor them. The one constraint was that the research be telecommunication-oriented. The output was knowledge possibly useful in the Bell system. Given this broad objective, the criteria of evaluation was excellent research that eventually might find its way into exploratory development. Through the exploratory development stage, the criteria were complete knowledge and excellent technology applied in the form a workable, economic system. The criteria at the development and production stages were compatibility with other systems, meeting schedules, cost savings, standardization, and efficiency. Implementation is the final stage of the technical logic, but it also was a criterion that pervaded the Bell organization as early as the discovery phase. When a system was approaching the implementation stage, however, we observed the importance of criteria like installation costs, reliability, and maintainability.

The primary criterion at NIH was "scientific merit" which emphasized methodological considerations or the theoretical relevance of the proposed research. The traditional evaluative mode was the peer review process which assessed new proposals. However, progress review often was left to the study sections, also, at the time a re-funding decision was made. From a results-oriented program viewpoint, this process may not permit timely adjustments either to changes in knowledge or to social pressures.

In the more successful NIH programs in our sample, the program managers assumed an active role in monitoring and evaluating projects. This was seen clearly in both the words and actions of the managers of the SIDS and genetics programs.

In the Sickle Cell Program, criteria for evaluating research or service projects never were established and evaluations made by the program coordinator generally were taken only as suggestions. The program coordinator made site visits to centers which had been established to link research and service at local levels and found mixed quality. One might have had an excellent research program and a poor community program, while another showed just the reverse characteristics. He suggested that centers with excellent science submit research program grant proposals, and those with a good community effort become clinics. However, the centers were supported through five-year grants, often viewed as gifts; although some centers were re-focused, the plan never became a practice.

Perhaps one of the significant reasons evaluation never became a part of the program was the lack of time. Everything moved so quickly that good administrative practices were pushed aside. The program coordinator commented:

> It was a crash effort. You couldn't plan it out. If I am unhappy about anything, it was not being able to, or not giving a lot of thought to establishing criteria for evaluation.

There also is a need for organizational or administrative criteria to link research programs to the larger organizational system or to environments beyond their boundaries. At the lower levels in Bell, there is more emphasis on technical matters than at higher levels where the concern is for project fit in a larger technical and organizational system. At various hierarchical levels, this latter concern was evident, but with different aspects emphasized and with different definitions of the "larger" system. The transmission cabinet would be concerned about how waveguide, for example, was progressing and how it continued to fit among the area's numerous transmission systems. At the next level up, the tri-company committee would focus on the Bell System in total and review waveguide from that perspective. The Bell System was tied together "horizontally" and "vertically" in a way best summarized by an earlier statement. Technical criteria govern the sequence of task activities, while the organizational criteria take cognizance of social, political, and market realities, and combine the evaluations in each area to determine the scale of programs.

There is evidence that a need for more integrative evaluation and criteria setting is recognized at NIH. In the past, it has been addressed by the individual program managers or by formation of trans-NIH committees for specific diseases. A satisfactory way of coordinating problems overlapping a number of institutes is still being sought. This type administrative endeavor generally is not an important consideration in the scientific research mind set and one can understand how it can be ignored. Although it can be a complicated and time-consuming task, our experience indicates that it is not one to be avoided.

Switching: The Process of Organizational Learning

Over time, organizations must adapt to survive. Sooner or later this calls for switching, the changing of important elements of the administrative process and the three inter-related logics. Changes may be reactive or proactive, seldom or frequent, minor or major, but they must be made. Some organizations clearly are more successful in carrying out this process than others and seem to have a greater capacity for it.

A source of change in knowledge generating organizations is the fact that task accomplishment, itself, necessitates shifts in the administrative process. When one phase of knowledge-building is largely finished, new realities call for a revised technical task logic and potentially a new set of administrative mechanisms. Similarly, if a program finds itself not making progress, it may need to reevaluate its technical logic and switch to a different stage in the innovation sequence as should have happened in the Artificial Heart Program. The successful programs we studied at NIH had this ability to shift activity to pick up knowledge necessary to achieve their goals. This appeared to be a localized phenomenon in that some program administrators possessed the intuitive ability to do it, while others, most notably in artificial heart, did not. At Bell, switching the focal stage of the technical logic appeared to be institutionalized and expected as part of achieving program success.

Organizations also have to cope with changes in their environments and may need to alter their political logic to adapt, as we saw happen in the SIDS Program. The logic of non-intervention into the affairs of the scientific community was broken and proactive outreach activities aimed at parents were initiated, as well.

A signal that it is time to switch seems to be the emergence of differences of opinion, changes in constraints such as resources, or other feedback which indicates difficulties with existing plans, schedules, or approaches. The wise response is to stop and resolve the issues and not to ignore them. These may be the only indications managers get suggesting that some shift is necessary. A less dramatic signal may be the absence of any conflict which, coupled with a lack of progress, may indicate a need to examine existing logics and redefine them.

These triggering conditions, requiring re-assessment of the ways research programs are approaching their technical and political environments, are closely related to the concept of variety. Once again, we can see evidence that, like tension, one can have too much, as well as too little, variety for the sake of learning. Figure 29 depicts this relationship and two examples of both kinds of errors. The original SIDS Program was bogged down technically with too little variety. It was the very explicit and deliberate infusion of fresh scientific viewpoints into the scene that broke the vicious cycle and got this program on to a fruitful learning track. The second example is one at the other extreme. An inability or unwillingness to modify the traditional, reactive, political logic of NIH contributed to a Sickle Cell Anemia Program being established and structured in a way that limited NIH's influence over its direction and activities. At the height of its volatile existence, the Sickle Cell Anemia Program clearly was swamped from the outside with diverse and powerful beliefs that could not be reconciled with its technically appropriate focus. The initial failure to change some elements of existing political and technical logics resulted in the program's being overwhelmed and never

Managing Large R & D Programs: Summary and Conclusions 151

developing its integrity. The scientific learning that should have contributed to the service activities was inhibited until the spotlight shifted elsewhere and the program settled down to calmer days.

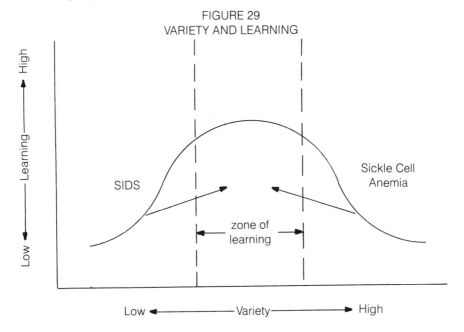

R & D executives monitor and guide a learning process that requires favorable conditions to proceed. This book is full of examples of this process becoming blocked in a variety of ways. The challenge to the R & D manager is to perceive these blockages and intervene to clear the blockage and reactivate learning.

In the course of this research, the authors came upon a wide variety of rules of thumb or heuristics that were usefully employed by program managers to introduce the conditions that induce learning. Even though it would be nice to have more elegant theories, with our limited knowledge, we must not snub any such rules. In the list below (Table 7) we have assembled a sample of these heuristics that struck us as useful guides to action in appropriate circumstances. We suggest that R & D executives take some time to "listen" to their organizations and determine whether their rules of thumb may inhibit or induce learning.

Using the Model

We would emphasize that our model clearly calls for the exercise of managerial judgment at every step of the way. It offers no promise of plugging

TABLE 7
SOME SELECTED HEURISTICS[1]

Condition	Heuristics
Creative variety:	—mix insiders and outsiders —mix old hands and newcomers —be skeptical of the experts
Creative tension:	—alternate organization bonds and barriers —give no single organization the power to go it alone —sell your ideas
Selective criteria:	—fund only the best —poor research results are only confusing and are worse than none at all —address emerging areas of need —use lean budgets —if it's not economical, it's not right —don't re-invent the wheel
Synthesis:	—use the hyphen (i.e., "targeted-basic") —utilize the one-system image —expressions of attitudes and beliefs associated with the dual advocacy leadership orientation.

[1]The work of Walter Buckley (1967) has influenced our thinking about organizations as learning systems. He identified variety, tension, and criteria as manifestations of adaptive systems. Heuristics, such as in the table, may be one way firms preserve and propagate successful adaptations. We think the possible link between language and the design of organizations may offer an intriguing approach to the study of organizational adaptation.

Managing Large R & D Programs: Summary and Conclusions 153

some raw data into a formula and cranking out a decision. Instead, it requires a manager to assess his or her own situation by addressing a series of questions that too often are left implicit in everyday affairs.

The first series of questions can be thought of as a check list—once answered, they lead to a second series of questions that call for judgments to be made about the quality of the fit or alignment between important aspects of the program. The third stage calls for drawing out implications from the analysis and initiating appropriate action steps. It probably would seem sensible to most senior R & D managers to start the review process by examining the technical logic of any program along with its associated administrative mechanisms. Any choice of a starting place is somewhat arbitrary. The most important rule is to keep moving to a completion of the analysis-action sequence. We would suggest starting the analysis with the less familiar and, therefore, more likely to be neglected area—the organization's political logic and its political environment.

The initial questions to be addressed are: Given a *tentative* formulation of program boundaries: 1) What are the more important groups with an interest in the program? 2) What logics exist for dealing with each of these groups? (non-intervention, intervention, etc.); 3) What are the key symbols in use? (normal professional relations, bothersome meddlers, routine reporting, etc.); 4) What organizational mechanisms and roles exist to deal with these groups? (advisory committees, liaison roles, periodic reports, newsletters, etc.); 5) In light of the above answers how would one characterize the pattern of mutual adaptation that is currently in use as perceived by the program? as perceived by the relevant external groups? (symbiosis or some less stable state such as domination, rejection, etc.); 6) What changes can be anticipated in these groups? in the research organization?

The answers to these questions can be put to use immediately in starting the second phase of identifying spots where there is a current or potential misalignment in the program's relations with its political environment. Here the focus probably would be on spotting the more severe actual or prospective misalignments. Any relationship that deserves a label other than that of symbiosis is, of course, a candidate for future, if not present, difficulties and such a characterization would lead into more detailed diagnostic questions. What factors, if changed, would upset a dominating relationship? What would have to be done to make it a more balanced exchange? What historical events contributed to a rejection state? What mechanisms and roles might open up a communication blockage? These are examples of the kinds of questions that tend to follow from the identification of a misalignment. A list of priority problem areas and possible lines of remedial action would be generated for management attention.

Most R & D managers probably feel on more familiar ground examining the organization's technical logic. The key requirement is to achieve an accurate reading on: 1) the dominant technical logic that is in use, 2) the state of knowledge of the physical phenomena, and 3) the relevant organizational mechanisms and roles that are in use. This, of course, is much easier said than done. The seven-stage model we have elaborated in this study should be helpful at this point. Is the technical logic of the organization in line with the state of knowledge or not? Is the array of mechanisms and roles appropriate for the stage of the technical logic?

This study and the bulk of previous research would support the rule of thumb that more organic mechanisms and roles align with the earlier stages of knowledge generation, and more formal and programmed mechanisms and roles align with the latter stages. Organic mechanisms and roles are characterized by a more open, face-to-face communication network, including horizontal and diagonal channels, by more participative decision making, by the deemphasis of status differentials and detailed role specifications. The more formal and programmed mechanisms and roles are characterized by constraints on communications outside the vertical authority channels, more authoritative decision making, more status differentials, and more reliance on specific role descriptions, detailed planning and scheduling, and formal performance measurement systems linked to formal rewards and sanctions. Any misalignments that are identified would be noted for re-examination and action.

Even after these two sets of diagnostic questions are answered, however, the R & D executive is still left without a very clear map as to how to translate his or her diagnosis into an action plan. We have attempted to provide some tentative assistance on this score by highlighting four areas of executive action earlier in this chapter. The final stage of program assessment leads back to a second look at the organization's political and technical logics. Do boundaries need to be changed? Are conflicts threatening program integrity? What steps would lead to their synthesis? Are existing criteria for evaluating technical output and political interactions adequate? Do identified misalignments require modification of political, technical, or organizational logics?

At this stage, all the pieces of the puzzle are on the table. Any misalignments within and among the three logics should have been highlighted. The question now becomes: How can the system be changed to move it closer to a complex, compound, moving alignment with its existing or anticipated future environment? Our model cannot answer this question fully. It only can require it being addressed in all its complexity rather than being ducked, and point out some areas in which to look for solutions. Trying to deal with all the complexity of research and development has helped us understand why programs can drift along for years without facing up to serious misalignments that eventually take their toll. But the more successful programs do have a record of addressing these fundamental, yet difficult, questions.

The model is straightforward, even obvious. As we began this research, however, we had no such guide, no framework for thinking about the issues we have discussed. Neither did NIH or Bell. These two prestigious and successful R & D institutions, staffed with talented people, experienced difficulty communicating with elements of their political environment about the nature of their tasks, and their organizational arrangements and management methods. Both believed a more systematic way of thinking and talking about large R & D programs was necessary. Our model is a step in that direction.

Bibliography

Abernathy, W. J. and Utterback J. M. "Innovation and the Evolving Structure of the Firm," Working Paper HBS 75–18, Harvard Business School, June 1975.
Allen, T. J. and Cohen, S. I. "Information Flow in Research and Development Laboratories," *Administrative Science Quarterly*, 1969.
Allen, T. J. "Communication Networks in R & D Laboratories", *R & D Management*, No. 1, 1970.
Allen, T. J. *Managing the Flow of Technology*, Cambridge: MIT Press, 1977.
Argyris, Chris A. *The Applicability of Organizational Sociology*, London: Cambridge University Press, 1972.
Argyris, Chris A. and Schon, Donald A. *Theory in Practice: Increasing Professional Effectiveness*, San Francisco: Jossey-Bass, 1974.
Ashby, W. R. *Design for A Brain*, London: Chapman and Hall Ltd., 1952.
Baker, W. O. "The Bell Labs/AT&T Relationship," *Bell Laboratories*, 1974.
Barnard, C. I. *The Functions of the Executive*, Cambridge, Massachusetts: Harvard University Press, 1938.
Barnes, Louis B. *Organizational Systems and Engineering Groups*, Boston, Massachusetts: Division of Research, Harvard Business School, 1960.
Barnowe, J. T. "Leadership and Scientific/Applied Outcomes in Research Organizations", Proceedings of the Second Annual Conference of the Canadian Association of Administrative Sciences, Toronto, Ontario, June 1974.
Barnowe, J. T. "Leadership and Performance Outcomes in Research Organizations—The Supervisor of Scientists as a Source of Assistance", *Organizational Behavior and Human Performance*. Vol. 14 No. 2, Oct. 1975.
Beer, S. *Brain of the Firm*. New York: McGraw-Hill, 1972.
Bell, Daniel. *The Coming of Post-Industrial Society*, New York: Basic Books, 1973.
Berger, P. L. and Luckman T. *The Social Construction of Reality*, Garden City, New York: Doubleday & Company, 1966.
Bode, H. W. *Synergy: Technical Integration and Technological Innovation in the Bell System*, Murray Hill, New Jersey: Bell Telephone Laboratories, 1971.
Boulding, Kenneth E. *A Primer on Social Dynamics*, New York: The Free Press, 1970.
Braybrooke, D. and Lindblom C. E. *A Strategy of Decision*, New York: The Free Press, 1963.

Bronowski, J. *Science and Human Values,* New York: Julian Messner Inc., 1956.
Bronowski, J. *The Common Sense of Science,* Cambridge, Massachusetts: Harvard University Press, 1958.
Brooks, J. *Telephone,* New York: Harper & Row, 1975.
Buckley, Walter. *Sociology and Modern Systems Theory,* Englewood Cliffs, N. J.: Prentice-Hall Inc., 1967.
Burns, T. and Stalker, G. M. *The Management of Innovation,* London: Tavistock Publications, Ltd., 1961.
Carlsson, B., Keane P., Martin J. B., "R & D Organizations As Learning Systems", *Sloan Management Review*, Vol. 17, No. 3, Spring 1976.
Churchman, C. West, and Schainblatt, A. "The Researcher and the Manager: A Dialectic of Implementation", *Management Science,* Vol. 11, No. 4, February, 1965.
Clarke T. E. editor, *R & D Management Bibliography,* The Innovation Management Institute of Canada, Ottawa, Ontario, 1976.
Crozier, Michel. *The Bureaucratic Phenomenon,* Chicago: The University of Chicago Press, 1964.
Culliton, B. "The Route from Obscurity to Prominence", *Science*, October 13, 1972.
Daedalus, Journal of the American Academy of Arts and Sciences. "Science and Its Public: The Changing Relationship", Summer 1974.
de Solla Price, D. J. *Little Science, Big Science,* New York: Columbia University Press, 1963.
Dunn, E. S., *Economic and Social Development: A Process of Social Learning,* Baltimore: The Johns Hopkins Press, 1971.
Emery, F. E. and Trist, E. L. *Towards A Social Ecology* London: Plenum Press, 1973.
Engels, Frederick, *Dialectics of Nature,* New York: International Publishers, 1940.
Evan, W. M. "Superior-Subordinate Conflict in Research Organizations", *Administrative Science Quarterly,* June, 1965.
Friedlander, F. "Performance Orientation of Research Scientists", Unpublished paper, Case Western Reserve University, 1970.
Galbraith, Jay. *Designing Complex Organizations,* Reading, Massachusetts: Addison-Wesley, 1973.
George, Alexander L. "The Case for Multiple Advocacy in Making Foreign Policy", *The American Political Science Review,* Vol. 66, 1972.
Glaser, B. G. "Differential Association and Institutional Motivation in Scientists", *Administrative Science Quarterly,* June 1965.
Glass, Bentley. *Science and Ethical Values,* Chapel Hill, N.C.: University of No. Carolina Press, 1965.
Gordon, Gerald. "Preconceptions and Reconceptions In the Administration of Science", in *Research Program Effectiveness,* Marshall C. Yovits, et. al., editors, New York; Gordon and Breach, 1966.
Gordon, G., et al. "A Contingency Model for the Design of Problem-Solving Research Programs: A Perspective on Diffusion Research", *Health and Society,* Spring 1974.

Gordon, G. and Fisher, G. L. *The Diffusion of Medical Technology: Policy and Planning Perspectives*, Cambridge, Massachusetts: Ballinger Publishing Co., 1975.

Gouldner, A. W. "Cosmopolitans and Locals: Toward An Analysis of Latent Social Roles", Administrative Science Quarterly, Vol. 2, 1951.

Gouldner A. W., *Patterns of Industrial Bureaucracy,* New York: Free Press, 1954.

Hower, Ralph M., and Orth, Charles D. *Managers and Scientists*, Boston, Massachusetts: Division of Research, Harvard Business School, 1963.

Jermakowicz W. "Organizational Structures in the R & D Sphere", *R & D Management*, Vol. 8 Special Issue on the Management of Research, Development and Education, 1978.

Keller, R. T. and Holland, W. E. "Boundary Spanning Roles in a Research and Development Organization: An Empirical Investigation", *Academy of Management Journal,* June 1975.

Kidd, Charles V. "Basic Research: Description Versus Definition", *Science*, Vol. 129, February, 1959.

Kolb D. A., "On Management and the Learning Process", Sloan School Working Paper 652—73, Cambridge, Massachusetts Institute of Technology, 1973.

Kornhauser, W., with W. O. Hagstrom. *Scientists in Industry: Conflict and Accommodation,* Berkeley: University of California Press, 1962.

Kotter, John P. and Lawrence Paul R. *Mayors in Action,* New York: John Wiley & Sons, 1974.

Kuhn, Thomas S. *The Structure of Scientific Revolutions*, Chicago: The University of Chicago Press, 1962.

La Porte, Todd R. "Conditions of Strain and Accommodation in Industrial Research Organizations", *Administrative Science Quarterly,* June 1965.

Lawrence P. and Lorsch, Jay W. *Organization and Environment*, Boston: Division of Research, Harvard Business School, 1967.

Lourenco, Susan V. and Glidewell, John C. "A Dialectical Analysis of Organizational Conflict", *Administrative Science Quarterly,* Vol. 20, December, 1975. pp. 489—508.

Mansfield, E. et al. *Research and Innovation in the Modern Corporation,* New York: Norton & Co , 1971.

March, James G. and Simon, Herbert A. *Organizations*, New York: John Wiley & Sons Inc., 1958.

Maslow, Abraham H. *The Psychology of Science: A Reconnaissance,* New York: Harper & Row, 1966.

Maslow, Abraham H. *New Knowledge in Human Values*, New York: Harper & Row, 1959.

Merton, Robert K. *The Sociology of Science,* Chicago: University of Chicago Press, 1973.

Merton, Robert K. *Social Theory and Social Structure*, New Yirk: The Free Press, 1968.

Miles, R. H. "How Job Conflicts and Ambiguity Affect R & D Professionals", *Research Management,* July, 1975.

Morton, Jack A. "From Research to Technology", *International Science and Technology*, May 1964, pp. 82—92.

Morton, Jack A. "A Systems Approach to the Innovation Process", *Business Horizons*, Summer 1967.
Morton, Jack A. *Organizing for Innovation: A Systems Approach to Technical Management*, New York: McGraw-Hill, 1971.
National Institute of General Medical Science. *Annual Reports* 1968–1975.
National Institutes of Health Almanac, U.S. Department of Health, Education and Welfare, Washington, 1975.
Pelz, D. C. "Social Factors Related to Performance in Research Organizations", *Administrative Science Quarterly*, December 1956.
Pelz, D. C. and Andrews, F. M. *Scientists in Organizations: Productive Climates for Research and Development*, New York: John Wiley and Sons, 1966.
Piaget, Jean, *Structuralism*, New York: Harper & Row, 1970.
Polanyi, M. *The Tacit Dimension*, New York: Doubleday, 1966.
Revelle, R. "The Scientist and the Politician", *Science*, Vol. 187, March 1975.
Rhenman, Eric. *Organization Theory for Long-Range Planning*, London: Wiley, 1973.
Ritchie, E. "Research on Research: Where Do We Stand?", *R & D Management*, 1971.
Roberts, Marc. "On the Nature and Condition of Social Science", *Daedalus*, Summer, 1974.
Robertson, A. B., et al. (Project SAPPHO) *Success and Failure in Industrial Innovation*, University of Sussex, 1972.
Robertson, A. B. *The Lesson of Failure*, London: MacDonald, 1974.
Roethlisberger, F. J. *The Elusive Phenomena*, edited by G. Lombard, Cambridge: Harvard University Press, 1977.
Rosenbloom, Richard S. "Technological Innovation in Firms and Industries. An Assessment of the State of the Art", Working Paper, Harvard Business School, 1974.
Rosenbloom, R. S. and Wolek, F. W. *Technology and Information Transfer: A Survey of Practice in Industrial Organizations*, Boston: Division of Research, Harvard Business School, 1970.
Satrow, R. L. "Value-Rational Authority and Professional Organizations: Weber's Missing Type", *Administrative Science Quarterly*, December, 1975.
Sayles, L. R. and Chandler, W. K. *Managing Large Systems*, New York: Harper & Row, 1971.
Schweitzer, E. "Forty Years of Waveguides: A Glimpse at History", *Bell Laboratories Record*, Vol. 48, No. 3, March, 1970.
Seiler, Robert E. *Improving the Effectiveness of Research and Development*, New York: McGraw-Hill, 1965.
Selznick, Phillip. *Leadership in Administration*, New York: Harper & Row, 1957.
Shepard, H. A. "Nine Dilemmas in Industrial Research", *Administrative Science Quarterly*; December, 1956.
Silverman, David. *The Theory of Organizations*, New York: Basic Books, 1971.
Simon, Herbert A. *The Sciences of the Artificial*, Cambridge, Massachusetts: The MIT Press, 1969.
Snow, C. P. *Science and Government*, Cambridge, Massachusetts: Harvard University Press, 1961.
Snow, C. P. *The Two Cultures and A Second Look*, Cambridge University Press,

Bibliography 161

1964.

Southworth, G. C. "Electric Wave Guides", *Bell Laboratories Record*, Vol. 14, No. 9, May, 1936.

Southworth, G. C. "Hyper-Frequency Wave Guides—General Considerations and Experimental Results", *Bell System Technical Journal*, March, 1936.

Steele, Lowell W. *Innovation in Big Business*, New York: American Elsevier, 1975.

Stetten, D., Shannon, J., Handler, P., Thomas, L., and Anderson, N. *On the Stewardship of Basic Science and the National Institute of General Medical Sciences*, remarks on the tenth anniversary of NIGMS, 1973.

Strickland, S. *Politics, Science and Dread Disease*, Cambridge: Harvard University Press, 1972.

Tagiuri, R. "Value Orientations and the Relationship of Managers and Scientists", *Administrative Science Quarterly*, June, 1965.

Tagiuri, R. and Guth, W. "Personal Values and Corporate Strategy", *Harvard Business Review*, September-October, 1965.

Thomas, Lewis. *The Lives of A Cell*, Viking Press, 1976.

Thompson, James D. *Organizations in Action*, New York: McGraw-Hill, 1967.

Twiss, B. C. *Managing Technological Innovation*, London: Longman Group, 1974.

U.S. Department of Health, Education and Welfare. *What Are the Facts About Genetic Disease?* DHEW Publication Number (NIH) 75-370, 1975.

U.S. Department of Health, Education and Welfare. *Forward Plan for Health 1977-1981*, Washington: Government Printing Office, 1975.

U.S. House of Representatives. Testimony Before the House Appropriations and Committee on Public Health and Environment, June 10, 1970.

U.S. House of Representatives. *Research Treatment and Prevention of Sickle Cell Anemia*, Hearings before Sub-committee on Public Health and Environment (of the Committee on Interstate and Foreign Commerce), 92nd Congress, November 12, 1971, Washington: Government Printing Office.

U.S. House of Representatives. *Sudden Infant Death Syndrome*, Hearings before Sub-committee on Public Health and Environment, 93rd Congress, August 2, 1973, Washington: Government Printing Office.

U.S. House of Representatives. *Heart, Lung and Blood Research Training and Genetic Disease Amendments of 1975*. Report by the Committee on Interstate and Foreign Commerce, 94th Congress, September 22, 1975, Washington: Government Printing Office, 1975.

U.S. Senate. *National Sickle Cell Anemia Prevention Act*. Hearings before the Subcommittee on Health (of the Committee on Labor and Public Welfare), 92nd Congress, November 11 and 12, 1971, Washington: Government Printing Office, 1972.

U.S. Senate. *Rights of Children 1972*, Hearings before Sub-committee on Labor and Public Welfare, 92nd Congress, Part 1, "Examination of the Sudden Infant Death Syndrome", January 25, 1972, Washington: Government Printing Office, 1972.

U.S. Senate. *National Biomedical Heart, Lung, Blood, Blood Vessel and Research Training Act of 1975*, Hearings before the Sub-committee of Health (of the Committee on Labor and Public Welfare), 94th Congress, March 17, 1975, Washington: Government Printing Office, 1975.

Utterback, J. M. "The Process of Innovation: A Study of the Origination and Development of Ideas for New Scientific Instruments", *IEEE Transactions on Engineering Management*. Vol. EM−18 No. 4, 1971.

Utterback, James M. "Innovation in Industry and the Diffusion of Technology", *Science*, February 15, 1974.

Vickers, Sir Geoffery. *The Art of Judgment*, New York: Basic Books, 1965.

Vickers, Sir Geoffery. *Value Systems and Social Processes*, New York: Basic Books, 1968.

Vickers, Sir Geoffery. *Making Institutions Work,* New York: John Wiley & Sons, 1973.

von Bertalanffy, Ludwig. *General Systems Theory: Foundations, Development, Applications*, New York: George Braziller, 1968.

Vygotsky, Lev S. *Thought and Language*, Cambridge, Massachusetts: The MIT Press, 1962.

Warters, W. D. "Millimeter Wave Guide Scores High in Field Test", *Bell Laboratories Record,* November, 1975.

Watson, James D. *The Double Helix,* New York: Mentor Books, 1968.

Weick, K. *The Social Psychology of Organizing,* Reading, Massachusetts: Addison-Wesley, 1969.

Weinberg, A. M. *Reflections on Big Science*, London: Pergamon Press, 1967.

Woodward, J. *Management and Technology*, London: H.M.S.O., 1958.

Woolridge Committee. *Bio-Medical Research and Its Administration*, Report of Presidential Committee, Washington: Department of Health, Education and Welfare, 1965.

Zaltman, G., Duncan, and Holbec J. *Innovation and Organization*, New York: Wiley, 1973.

Index

Adaptation, organizational: boundaries and, 142–44; difficulties encountered in, 122–23; four factors in, 141; of Genetics Program, 129–31, 136; management criteria and, 148–49; patterns of, 29–32; switching and, 149–51; synthesis and, 145–47; two modes of, 138 (fig.)
Administrative mechanisms, 89–97, 147, 154
Allen, T. J. (1972), 3
American Cancer Society, 66, 68, 69
Applied research. *See* Basic research
Argyris, Chris (1972), 63
Artificial Heart Program, boundary management in, 143; compared to Cancer Chemotherapy Program, 53, 72; contrasted with Millimeter Waveguide Program, 63–66, 64–65 (fig.); contrasted to SIDS and Sickle Cell Anemia Programs, 86–87; historical background, 54–58; political and technical logic of, 122–23; political context of, 59–60; reasons for selecting, 12; task logic of, 71 (fig.); technical logic of, 60–63; technique orientation of, 94–95
AT&T, 6, 28, 106, 147, 148, 149; administrative mechanisms, 90–91, 94; boundary management in, 143–44; emphasis on development, 36, 108–10; general contrast with NIH, 8–9; involvement in No. 4 ESS program, 114, 115, 117, 118 (fig.); organizational logic at, 102–3, 103 (fig.), 119–21. *See also* Bell Telephone Laboratories; Long Lines Department; Western Electric
Auto industry, political intervention in, 6

Barnowe, J. T. (1975), 3
Basic research, 104; vs. applied research, 7, 34–35, 125–26, 129–30, 132

Bell Telephone Laboratories (Bell Labs), 8–9, 41, 142; branch labs at Western Electric, 110–12; and D4 Program, 112–14; and HCMTS Program, 27–28, 29; and Millimeter Waveguide Program, 46, 47–48; and No. 4 ESS Program, 114–16, 117; organization of, 100–103, 101 (fig.). *See also* Crawford Hill
Bell System. *See* AT&T
Bergman, Fred, 131–32, 133–34, 134–35
Biomedical research, 38–40, 40 (table); compared to communications research, 41–44, 50
Black box, approach to research, 38–39, 61, 63, 68
Black community, 19–20, 21, 24
Boundary management, 134–37, 136 (fig.), 141–44
Buckley, Walter, 152n

Cancer Chemotherapy National Service Center (CCNSC), 68
Cancer Chemotherapy Program, 53, 90, 91, 144; background, 67–69; political and technical logic of, 69–74, 70 (fig.), 72 (fig.); reasons for selecting, 12
Clinical trials, 39
Communications research, 41–44, 41 (fig.)
Computer systems, 117–19
Congress, U.S., and Artificial Heart Program, 57–58; and Cancer Chemotherapy Program, 68, 69, 73–74; and Genetics Program, 124, 129–30; and Sickle Cell Anemia Program, 19, 21–23; and SIDS Program, 81–83
Contracts, vs. grants, 90, 129–30
Cooper Committee, 131
Crawford Hill, 104–8

163

Index

Defense, Department of, 56
D4 Digital Channel Bank Program, 12–13, 112–14
De Solla Price, D. J. (1963), 30n
Development (phase of research), 35, 47–49, 109–10
Division of Cancer Treatment (DCT), 68–69
Dole, Robert J., 74
Division of General Medical Sciences, 124
Dual advocacy, 146–47

Economic context (environment), 4–5. *See also* R & D funding
Empirical research logic, 42–44, 43 (fig.), 71; in Cancer Chemotherapy Program, 91
Engineering Research Center (ERC), 47–48
Exploratory development (phase of research), 108–9

Federal Communications Commission (FCC), 25–28, 29
Fogarty, John, 67, 124, 135
Fredrickson, Donald, 74
Friedlander, F. (1970), 3
Funding, research and development. *See* R&D funding

Genetic counseling research, 133
Genetics, 36–38
Genetics Program, adaptation to environment of, 123, 127 (fig.), 128; after 1972, 131–37; before 1972, 124–26; dual advocacy in, 146–47; organizational logic of, 137 (fig.); reasons for selecting, 12
Goldberg, Mr. and Mrs. Saul, 77
Gouldner, A. W. (1951), 3
Grants, vs. contracts, 90, 129–30

Hasselmeyer, Eileen, development of SIDS Program, 79–81; management of parent groups, 83–84; personal leadership in SIDS Program, 86, 87–89; responses to concerns of Congress, 81–83; use of OPS by, 84–85, 91
Hastings, Frank, 94–95
Health, Education and Welfare, Department of (HEW), 6–8, 21, 84
Health Services and Mental Health Administration, 21
Heuristics, 152 (fig.)
High Capacity Mobile Telephone System (HCMTS) Program, 13, 25–28, 32, 108–9
Hill, Lister, 67, 124, 135
Hunter, Jehu, 79–81, 83–84

Innovation, 5, 35–36, 43
International Guild for Infant Survival, 77

Jermakowicz, W. (1978), 4
Johnson, Lyndon B., 79

Kornhauser, W., and Hagstrom, W. O. (1962), 3
Kotter, John P., and Lawrence, Paul R. (1974), 147

Lasker, Albert and Mary, 66–67
Leadership, 3, 94–97
Lederberg, Joshua, 55
Linkage, critical, 50–51, 63
Logic. *See* Organizational logic; political logic; technical logic
Long Lines Department (AT&T), 46, 48–49, 119
Loop, learning, 43–44, 51–52, 62, 98 (fig.)
Lourenco, Susan V., and Glidewell, John C. (1975), 145

McGovern, George, 74
Management, research and development. *See* R&D management
Mansfield, E., et al. (1971), 4, 102
Mark Addison Roe Foundation, 77
Merrimac Valley Works. *See* Western Electric
Merton, Robert K. (1968), 5, 129n
Merton, Robert K. (1972), 5
Millimeter Waveguide Program, administrative review of, 94; compared to Artificial Heart Program, 63–66, 64–65 (fig.); development approach to, 108; history of, 44–52; political and technical logic in, 123; program orientation of, 95–97; reasons for selecting, 13, technical success of, 52
Model, R&D management, 139, 140–42, 141 (fig.), 151–55
Morton, Jack, 110
Myocardial Infarction Program, 58

NASA, 55, 56, 60–61
National Association of Radio Systems (NARS), 25, 27
National Cancer Institute, 11, 66–67, 68
National Foundation for Sudden Infant Death Syndrome, 77
National Heart and Lung Institute (NHLI), 20–21, 54–55
National Institute of Child Health and Human Development (NICHD), 20, 77, 79, 81–83
National Institute of General Medical

Index

Sciences,(NIGMS), 20, 124-125, 126, 129-131
National Institutes of Health (NIH), attitude toward grants and contracts in, 130; central administration and Artificial Heart Program, 57-58, 87; central administration and NIGMS, 125, 131; central administration and Sickle Cell Anemia Program, 87; central administration and SIDS Program, 84-85; contrasted to AT&T, 8-9, 121; evaluative criteria at, 148, 149; funding for, 6-8, 82, 90, 135; and grant applications, 126; noninterventionist role, 82; as political context of individual programs, 59-60; political intervention in, 8, 14; polycentrism of, 59-60, 62, 97; reasons for study of, 6-8; scientific orientation of, 36; structure and boundary management by, 143. *See* separate listings for individual institutes and programs
New York Times, on Cancer Chemotherapy Program, 73-74
Nixon, Richard M., 19, 131
Number 4 Electronic Switching System (No. 4 ESS) Program, 13, 114-19

Office of Management and Budget (OMB), 6-8
Operational Planning System (OPS), 84-85, 86, 91, 92-93 (fig.)
Organizational logic, boundary management in, 142-44; criteria in, 147-49; defined, 15; in Genetics Program, 126, 131-36, 137 (fig.); and knowledge transfer, 107-9, 110, 113-14 (figs.); lessons learned from AT&T, 119-21; a model for, 140-42, 141 (fig.), 151-55; of No. 4 ESS Program, 114-19; one-system concept of, 120-21, 121 (fig.); switching in, 149-51; synthesis in, 145-47. *See also* Political logic; technical logic
Organizational space, 15
Organizational structure, of AT&T, 103 (fig.); of Bell Labs, 100-103, 101 (fig.); at Crawford Hill, 104-8; of No. 4 ESS Program, 118 (fig.)

Packwood, Robert, 79
Parents' organizations, 77, 81-82, 85-86
Pelz, D. C., and Andrews, F. M. (1966), 3
Penzias, Arno, 106
Perinatal Biology and Infant Mortality Branch (PBIM), 79, 80, 82, 85-86
PERT charts, 2, 89
Political context (environment), of Artificial Heart Program, 59-60; attitudes toward, 86-87; of Cancer Chemotherapy Program, 69; described, 16-18; dominance of, 53-54; of HCMTS Program, 28-29; in R&D management model, 140-41; of Sickle Cell Anemia Program, 23-25; of SIDS Program, 77-79, 81-86; and social context, 5-6; and technological core, 29-32
Political intervention in R&D, 5-6, 23-25, 33-34
Political logic, 15, 69-74, 123, 153. *See also* Organizational logic
Political space, 15
President's Biomedical Research Panel, 7
President's Commission on Heart Disease, Cancer, and Stroke, 56
Private research and development organizations, 31
Process, research and development. *See* R&D process
Product Engineering Control Center, 113-14, 114 (fig.)
Program, as unit of analysis, 13

Radio common carriers (RCCs), 25-29
R&D (research and development), focus of present study, 8-9; other studies on, 3-4; sample selection in present study, 9-12
R&D funding, 1-2, 91-92, 140; for Artificial Heart Program, 57-58, 62; for Cancer Chemotherapy Program, 53-54, 68; for Genetics Program, 133; for NIGMS, 125. *See also* Grants
R&D management, contrasting attitudes toward, 7-8; fads in, 2-3; heuristics for, 152 (fig.); model for, 139, 140-42, 141 (fig.), 151-55; and organizational adaptation, 122; other studies on, 3-5; and political context, 32; and technical logic, 33. *See also* Organizational logic
R&D process, in Artificial Heart Program, 61-62, 64-65 (fig.); in Millimeter Waveguide Program, 64-65 (fig.); stages in, 4, 34-35, 45 (fig.), 61, 63-66, 64-65 (fig.), 98 (fig.); and technical logic, 33; typologies for, 34-36
Research strategies, 68-69, 71, 72, 80-81. *See also* Technical logic

SAPPHO (1972), 5
Scheele, Leonard, 66
Schelkunoff, S. A., 44-46
Science, 30, 134-36
Sequential research logic, 42-44, 43 (fig.)
Shannon, James, 67
Shared Production Interactive Data Base for Error Reduction (SPIDER), 117-19

Sickle cell anemia, 18–19, 22
Sickle Cell Anemia Program,
  accomplishments of, 23; boundary
  management of, 142–43; contrasted to
  SIDS and Artificial Heart Programs,
  86–87; evaluative criteria in, 148–49;
  organizational adaption of, 122, 150–51;
  political context of, 19–23; political
  intervention in, 23–25; reasons for
  selecting, 12
Sloan Kettering Institute, 67
Social context (environment), 5, 6, 17,
  39–40; of Artificial Heart Program, 55; of
  Genetics Program, 125–26, 128, 134–36
Southworth, G. C., 44–46
Special interest groups, 17
Specialized Mobile Radio Systems (SMRs),
  27
Stetten, DeWitt, 130
Strickland, S. (1972), 67n
Sudden Infant Death Syndrome (SIDS)
  Program, boundary management by, 143;
  contrasted to Artificial Heart and Sickle
  Cell Anemia Programs, 86–87; described,
  76–77; dual advocacy in, 146; initiation
  of, 79–91; interactions with political
  environment by, 81–86; OPS chart for,
  92–93 (fig.); organizational adaptation of,
  150–51; and political and technical logic,
  123; political environment of, 77–79;
  reasons for selecting, 11; success of,
  75–76; technical logic of, 87–89
Symbiosis, 31–32, 76
Synthesis, organizational, 145–47
Systems-thinking, 62

Targeted research, 129–30
Technical context (environment), 53–78, 128
Technical logic, and administrative
  mechanisms, 89–97; of Artificial Heart
  Program, 60–63; of AT&T, 113 (fig.),
  121 (fig.); at Bell Labs, 100–121, 102
  (fig.); biological and communications
  research compared, 41–44; of biomedical
  research, 38–40; of Cancer Chemotherapy
  Program, 69–74; defined, 15; of Genetics
  Program, 132–34; and knowledge transfer,
  102–3; of Millimeter Waveguide Program,
  49–52; models of, 43 (fig.); and
  organization of No. 4 ESS Program, 116;
  and political logic, 123; in R&D
  management model, 153–54; of SIDS
  Program, 87–89; stages in, 38–44, 40
  (table); summarized, 140–41; typologies
  of, 34–36. See also Organizational logic
Technological core, 3, 6
Technological space, 14–15
Thompson, James D. (1967), 144, 147
Twiss, B. C. (1974), 4

Upton, Arthur C., 73, 74

Value-advocacy, 28
Value orientations, 3
Vaughn, Earle, 115, 116
Vickers, Geoffery (1968), 30
Vickers, Geoffery (1973), 28

Warters, Bill, 95–97, 120
Washington State Association for Sudden
  Infant Death Syndrome Study, 77
Washington State Legislature, 79
Waveguide theory, 44–46. See also
  Millimeter Waveguide Program
Weicker, Lowell, 77
Weinberg, A. M. (1967), 30
Western Electric, and D4 Digital Channel
  Bank Program, 112–14; interaction with
  Bell Labs, 110–12, 120; and No. 4 ESS
  Program, 117; and Millimeter Waveguide
  Program, 46, 47–48
Wireline companies, 25–29
World Health Organization, 76